Advances in Clinical Chemistry and Laboratory Medicine

Edited by Harald Renz and Rudolf Tauber

Advances in Clinical Chemistry and Laboratory Medicine

Edited by Harald Renz and Rudolf Tauber

DE GRUYTER

Editors

Prof. Dr. med. Harald Renz
Universitätsklinikum Giessen/Marburg
Klinische Chemie und Molekulare Diagnostik
Baldingerstraße 1
D-35043 Marburg
Germany

Prof. Dr. med. Rudolf Tauber
Charité-Universitätsmedizin Berlin
Institut für Laboratoriumsmedizin
Klinische Chemie und Pathobiochemie
Augustenburger Platz 1
D-13353 Berlin
Germany

The book contains 24 figures and 22 tables.

ISBN 978-3-11-022463-4
e-ISBN 978-3-11-022464-1

Library of Congress Cataloging-in-Publication data
A CIP catalog record for this book has been applied for at the Library of Congress.

Bibliographic information published by the Deutsche Nationalbibliothek
The Deutsche Nationalbibliothek lists this publication in the Deutsche Nationalbibliografie;
detailed bibliographic data are available in the Internet at http://dnb.d-nb.de.

Typesetting: Compuscript Ltd., Shannon, Ireland
Printing: Hubert & Co. GmbH & Co. KG, Göttingen

♾ Printed on acid-free paper
Printed in Germany
www.degruyter.com

Preface

Clinical chemistry and laboratory medicine are progressing at a rapid pace. Innovations in analytical and information technologies, e.g. genome sequencing, mass spectrometry and bioinformatics, have been and are being transferred from the research lab to the routine lab making new, powerful diagnostic tools available. At the same time, we are awaiting a much better understanding of the genomic, transcriptomic and metabolomic regulation with regard to the pathophysiology of diseases and, hence, the identification of new diagnostic biomarkers. These advancements will open new areas for laboratory diagnostic medicine, for instance patient risk assessment, prediction of disease development and evaluation of preventive strategies. Furthermore, the possibility of (sub) phenotyping with the introduction of individualized and tailored pharmacotherapy are already on the horizon.

At the same time, medicine is facing a number of challenges worldwide, in particular an increase in non-communicable diseases, such as chronic inflammatory conditions, auto-immunity, allergy, diabetes mellitus, obesity, cardiovascular disease, neurodigeneration and others. The prevalence of such conditions is certainly highest in industrialized countries; however, developing countries around the world are currently observing the same trend. Therefore, it is necessary for the results of medical research to be rapidly translated "from bench to bedside".

The fast rate of research and development makes it difficult to keep informed, particularly if one has in mind that laboratory medicine and clinical chemistry do not stand alone, but are closely linked with other fields in medicine and science, such as molecular biology, biochemistry, hematology, microbiology, immunology, transfusion medicine and other neighboring disciplines.

It is the aim of this volume to present a selection of articles that reflect recent advancements in the various areas of clinical chemistry and laboratory medicine. The papers were delivered at the IFCC WorldLab/EuroMedLab Congress held in Berlin, Germany, on May 15–19, 2011.

Harald Renz
Rudolf Tauber

Table of Contents

Authors Index

Khosrow Adeli
Department of Biochemistry
University of Toronto
Ontario
M5G 1X8
Canada
Department of Pediatric Laboratory Medicine
The Hospital for Sick Children
Toronto
Ontario M5G 1X8
Canada

Aleksandr A. Ammosov
The Clinical and Experimental Biochemistry
Laboratory
Federal V. Shumakov Research Center of
Transplantology and Artificial Organs
Moscow
Russia
transplant2009@mail.ru

Shereen H. Atef
Department of Clinical Pathology
Faculty of Medicine
Ain Shams University
Egypt
shereenatef@hotmail.com

William A. Bartlett
Blood Sciences Department
Diagnostics Group
Ninewells Hospital and Medical School
Dundee DD1 9SY
Scotland
UK
Bill.Bartlett@nhs.net

Gunther Becher
BioTeZ Berlin-Buch GmbH
Robert-Rössle-Strasse 10
13125 Berlin
Germany

Jörg Berg
Institute of Laboratory Medicine
General Hospital Linz
Krankenhausstrasse 9
A-4020 Linz
Austria
joerg.berg@akh.linz.at

Siegfried Bien
Abteilung für Neuroradiologie
Universitätsklinikum Giessen und Marburg
Rudolph-Bultmann-Straße 8
D-35033 Marburg
Germany

Saskia Brunner-Agten
Kantonsspital Aarau
Zentrum für Labormedizin
Tellstrasse
5001 Aarau
Switzerland

Henriett Butz
Second Department of Medicine
Faculty of Medicine
Semmelweis University
Budapest
Hungary
henriettbutz@gmail.com

Alberta Caleffi
U.O Diagnostica Ematochimica
Dipartimento Patologia e Medicina di
Laboratorio
Azienda Ospedaliero Universitaria
Parma
acaleffi@ao.pr.it

Massimiliano Cantinotti
Fondazione CMR – Regione Toscana G.
Monasterio
Ospedale del Cuore
Via Aurelia Sud
54100 Massa
Italy
Cantinotti@ifc.cnr.it, cantinotti@hotmail.it

Graham D. Carter
Imperial College Healthcare NHS Trust
Charing Cross Hospital
Fulham Palace Road
London W6 8RF
UK
b.carter1@which.net

Julie Chemin
Molecular Structure and Function
Research Institute
The Hospital for Sick Children
Toronto
Ontario M5G 1X8
Canada

Aldo Clerico
Fondazione G. Monasterio
Heart Hospital
Massa
Scuola Superiore Sant'Anna
Pisa
Italy

Anne H. Cross
Washington University School of Medicine
Campus Box 8111
660 Euclid Avenue S.
Saint Louis
Missouri
USA
crossa@neuro.wustl.edu

Sándor Czirják
National Institute of Neurosurgery
Budapest
Hungary

Alexandra Dorn-Beineke
Institut für Labordiagnostik und Hygiene
HSK
Dr. Horst Schmidt Kliniken GmbH
Ludwig-Erhard-Straße 100
D-65199 Wiesbaden
Germany

Henning Ebelt
Department of Medicine III
University Clinics Halle (Saale)
Martin-Luther-University Halle-Wittenberg
Germany

Mohammed E. Elfaki
The Leishmaniasis Research Group/Sudan
The Institute of Endemic Diseases
Medical sciences campus
University of Khartoum
P.O. Box 102
Khartoum
Sudan

Ahmed M. El-Hassan
The Leishmaniasis Research Group/Sudan
The Institute of Endemic Diseases
Medical sciences campus
University of Khartoum
P.O. Box 102
Khartoum
Sudan
nahnoh80@hotmail.com, eltahirk@iend.org,
eltahirgasim@yahoo.ca

Nuha A.A. Elnojomi
The Leishmaniasis Research Group/Sudan
The Institute of Endemic Diseases
Medical sciences campus
University of Khartoum
P.O. Box 102
Khartoum
Sudan

Amr Fattouh
Department of Clinical Pathology
Faculty of Medicine
Ain Shams University
Egypt

Abd El Rahman Gaber
Department of Ophthalmology
Faculty of Medicine
Ain Shams University
Egypt

Stepan Gambaryan
Institut für Klinische Biochemie and
Pathobiochemie – Zentrallabor (IKBZ)
Zentrum Innere Medizin
Universitätsklinikum
Oberdürrbacher Str. 6 – Haus A4
97080 Würzburg
Germany

David Gancberg
IFCC Committee for Molecular Diagnostics

Holger Garn
Philipps University Marburg
Institute of Laboratory Medicine and
Pathobiochemistry
Molecular Diagnostics
Baldinger Straße
35043 Marburg
Germany

Sergey V. Gautier
Department of Clinical Transplantology
Federal V. Shumakov Research Center of
Transplantology and Artificial Organs
Moscow
Russia
transplant2009@mail.ru

Maria Geit
Department of Dermatology
Institute of Laboratory Medicine
General Hospital Linz
Krankenhausstrasse 9
A-4020 Linz
Austria

Wolfram Hubert Gerlich
Institute for Medical Virology
Justus-Liebig-University Giessen
Frankfurter Str. 107
35392 Giessen
Germany
Wolfram.H.Gerlich@viro.med.uni-giessen.de

Olga E. Gichkun
Department of Clinical Transplantology
Federal V. Shumakov Research Center of
Transplantology and Artificial Organs
Moscow
Russia
transplant2009@mail.ru

Dieter Glebe
Institute for Medical Virology
Justus-Liebig-University Giessen
Frankfurter Str. 107
35392 Giessen
Germany

Thomas Häupl
BioTeZ Berlin-Buch GmbH
Robert-Rössle-Strasse 10
13125 Berlin
Germany
info@biotez.de

Albert J.R. Heck
Netherlands Proteomics Centre
Biomolecular Mass Spectrometry and
Proteomics Group
Utrecht University
Padualaan 8
3584 CH Utrecht
The Netherlands
a.j.r.heck@uu.nl

Wolfgang Herrmann
Saarland University Hospital
Department of Clinical Chemistry and
Laboratory Medicine
Central Laboratory
Building 57
66421 Homburg/Saar
Germany

Sabine Herterich
Institut für Klinische Biochemie and
Pathobiochemie – Zentrallabor (IKBZ)
Zentrum Innere Medizin
Universitätsklinikum
Oberdürrbacher Str. 6 – Haus A4
97080 Würzburg
Germany

Andreas R. Huber
Kantonsspital Aarau
Zentrum für Labormedizin
Tellstrasse
5001 Aarau
Switzerland
Andreas.huber@ksa.ch

Wim Huisman
Medical Centre Haaglanden
P.O. BOX 432
2501 CK Den Haag
The Netherlands
w.huisman@mchaaglanden.nl

Péter Igaz
Second Department of Medicine
Faculty of Medicine
Semmelweis University
Budapest
Hungary

Boris Ivandic
DIAneering – Diagnostics Engineering and
Research GmbH
Heidelberg
Germany

Glenville Jones
Craine Professor of Biochemistry and
Professor of Medicine
Department of Biomedical and Molecular
Sciences
Queen's University
Room 651
Botterell Hall
Kingston
Ontario K7L 3N6
Canada
gj1@queensu.ca

Graham Jones
Department of Chemical Pathology
St Vincent's Hospital
Victoria St
Darlinghurst
2010 NSW
Australia
gjones@stvincents.com.au

Martha Kaeslin
Kantonsspital Aarau
Zentrum für Labormedizin
Tellstrasse
5001 Aarau
Switzerland

Eltahir A. Khalil
The Leishmaniasis Research Group/Sudan
The Institute of Endemic Diseases
Medical sciences campus
University of Khartoum
P.O. Box 102
Khartoum
Sudan

May-Jean King
Membrane Biochemistry
International Blood Group Reference
Laboratory
NHS Blood and Transplant
500 North Bristol Park
Northway
Filton
Bristol BS34 7QH
UK
may-jean.king@nhsbt.nhs.uk

Susanne H. Kirsch
Saarland University Hospital
Department of Clinical Chemistry and
Laboratory Medicine
Central Laboratory
Building 57
66421 Homburg/Saar
Germany
su.kirsch@uniklinikum-saarland.de

Tilmann Otto Kleine
Institut für Laboratoriumsmedizin und
Pathobiochemie
Molekulare Diagnostik
Universitätsklinikum Giessen und Marburg
Referenzlabor für Liquordiagnostik
Baldingerstraße
D-35033 Marburg
Germany
kleine@staff.uni-marburg.de

Petr Kocna
Institute of Clinical Biochemistry and
Laboratory Diagnostics
First Medical Faculty
Charles University
Karlovo namesti 21
CZ-121-11
Prague 2
Czech Republic
kocna@lf1.cuni.cz

Márta Korbonits
Department of Endocrinology
Barts and the London School of Medicine
Queen Mary University of London
London EC1M 6BQ
UK

Cara S. Kosack
Médecins Sans Frontières
Plantage Middenlaan 14
1018 DD Amsterdam
The Netherlands
cara.kosack@amsterdam.msf.org

Rivada M. Kurabekova
The Clinical and Experimental Biochemistry
Laboratory
Federal V. Shumakov Research Center of
Transplantology and Artificial Organs
Moscow
Russia
transplant2009@mail.ru

Reinhardt Lehmitz
Neurologische Klinik
Zentrallabor für Liquordiagnostik
Universität Rostock
D-18147 Rostock
Germany

István Likó
Gedeon Richter Plc
Budapest
Hungary

Suzanne M. Lohmann
Institut für Klinische Biochemie and
Pathobiochemie – Zentrallabor (IKBZ)
Zentrum Innere Medizin
Universitätsklinikum
Oberdürrbacher Str. 6 – Haus A4
97080 Würzburg
Germany

Christa Löwer
Institut für Laboratoriumsmedizin und
Pathobiochemie
Molekulare Diagnostik
Universitätsklinikum Giessen und Marburg
Referenzlabor für Liquordiagnostik
Baldingerstraße
D-35033 Marburg
Germany

Nermine H. Mahmoud
Department of Clinical Pathology
Faculty of Medicine
Ain Shams University
Egypt

Cyril D.S. Mamotte
IFCC Committee for Molecular Diagnostics

Eva-Maria Matzhold
University Clinic of Blood Group Serology and
Transfusion Medicine
University Hospital Graz
Medical University Graz
Graz
Austria

Hans H. Maurer
Department of Experimental and Clinical
Toxicology
Saarland University
66421 Homburg (Saar)
Germany
hans.maurer@uks.eu

Gabriel Alejandro Migliarino
Gmigliarino Consultants
Rodriguez Peña 1054
Castelar
Buenos Aires CP 1712
Argentina
gmigliarino@gmigliarino.com

Bruno Murzi
Fondazione G. Monasterio
Heart Hospital
Massa
Scuola Superiore Sant'Anna
Pisa
Italy

Ahmed M. Musa
The Leishmaniasis Research Group/Sudan
The Institute of Endemic Diseases
Medical sciences campus
University of Khartoum
P.O. Box 102
Khartoum
Sudan

Peter Nollau
Institute of Clinical Chemistry
University Medical Center Hamburg-
Eppendorf
Martinistr. 52
20246 Hamburg
Germany

Rima Obeid
Saarland University Hospital
Department of Clinical Chemistry and
Laboratory Medicine
Central Laboratory
Building 57
66421 Homburg/Saar
Germany

Christian Paar
Institute of Laboratory Medicine
General Hospital Linz
Krankenhausstrasse 9
A-4020 Linz
Austria

Attila Patócs
Molecular Medicine Research Group
Hungarian Academy of Sciences
Second Department of Medicine
Faculty of Medicine
Semmelweis University
Budapest
Hungary

Deborah A. Payne
American Pathology Partners UniPath
Molecular Services
6116 East Warren Ave
Denver
Colorado 80222
USA
IFCC Committee for Molecular Diagnostics
dpayne@unipathdx.com

Mario Pazzagli
IFCC Committee for Molecular Diagnostics

Petra Ina Pfefferle
Philipps University Marburg
Institute of Laboratory Medicine and
Pathobiochemistry
Molecular Diagnostics
Baldinger Straße
35043 Marburg
Germany

Károly Rácz
Second Department of Medicine
Faculty of Medicine
Semmelweis University
Budapest
Hungary

Jens G. Reich
BioTeZ Berlin-Buch GmbH
Robert-Rössle-Strasse 10
13125 Berlin
Germany
info@biotez.de

Harald Renz
Philipps University Marburg
Institute of Laboratory Medicine and
Pathobiochemistry
Molecular Diagnostics
Baldinger Straße
35043 Marburg
Germany
renzh@med.uni-marburg.de

Dirk Roggenbuck
GA Generic Assays GmbH
Ludwig-Erhard-Ring 3
15827 Dahlewitz
Germany
dirk.roggenbuck@genericassays.com

François Rousseau
IFCC Committee for Molecular Diagnostics

Juan Manuel Acedo Sanz
Hospital Universitario Fundación de Alcorcón
Avenida de Budapest s/n
Alcorcón
Madrid
España
juanma.acedo@gmail.com

Dieter Sarrach
BioTeZ Berlin-Buch GmbH
Robert-Rössle-Strasse 10
13125 Berlin
Germany
info@biotez.de

Ron H.N. van Schaik
IFCC Committee for Molecular Diagnostics

Astrid Schäfer
BioTeZ Berlin-Buch GmbH
Robert-Rössle-Strasse 10
13125 Berlin
Germany

Heinz Schimmel
IFCC Committee for Molecular Diagnostics

Arjen Scholten
Netherlands Proteomics Centre
Biomolecular Mass Spectrometry and
Proteomics Group
Utrecht University
Padualaan 8
3584 CH Utrecht
The Netherlands
a.scholten@uu.nl

Christian Gisbert Schüttler
Institute for Medical Virology
Justus-Liebig-University Giessen
Frankfurter Str. 107
35392 Giessen
Germany

Olga P. Shevchenko
The Clinical and Experimental Biochemistry
Laboratory
Federal V. Shumakov Research Center of
Transplantology and Artificial Organs
Moscow
Russia
transplant2009@mail.ru

Eberhard Spanuth
DIAneering – Diagnostics Engineering and
Research GmbH
Heidelberg
Germany
eberhard.spanuth@t-online.de

Luděk Šprongl
Central laboratory
Šumperk Hospital
Nerudova 640/41
78752 Šumperk
Czech Republic
sprongl@nemspk.cz

Antonia Staatz
BioTeZ Berlin-Buch GmbH
Robert-Rössle-Strasse 10
13125 Berlin
Germany
info@biotez.de
info@biotez.de

Jens-Oliver Steiß
BioTeZ Berlin-Buch GmbH
Robert-Rössle-Strasse 10
13125 Berlin
Germany
info@biotez.de

Herbert Stekel
Institute of Laboratory Medicine
General Hospital Linz
Krankenhausstrasse 9
A-4020 Linz
Austria

Simona Storti
Fondazione G. Monasterio
Heart Hospital
Massa
Scuola Superiore Sant'Anna
Pisa
Italy

Michael Stowasser
Endocrine Hypertension Research Centre
University of Queensland School of Medicine
Greenslopes and Princess Alexandra Hospitals
Brisbane
Australia
m.stowasser@uq.edu.au

Pavel Strohner
BioTeZ Berlin-Buch GmbH
Robert-Rössle-Strasse 10
13125 Berlin
Germany
info@biotez.de

Olga M. Tsirulnikova
The Clinical and Experimental Biochemistry
Laboratory
Federal V. Shumakov Research Center of
Transplantology and Artificial Organs
Moscow
Russia
transplant2009@mail.ru

Donald R.A. Uges
Laboratory for Therapeutic Drug Monitoring
Clinical and Forensic Toxicology
Department of Hospital Pharmacy
University Medical Center Groningen
P.O. Box 30.001
9700 RG Groningen
The Netherlands
d.r.a.uges@apoth.umcg.nl

Christoph Wagener
Institute of Clinical Chemistry
University Medical Center Hamburg-
Eppendorf
Martinistr. 52
20246 Hamburg
Germany
wagener@uke.de

Ulrich Walter
Institut für Klinische Biochemie and
Pathobiochemie – Zentrallabor (IKBZ)
Zentrum Innere Medizin
Universitätsklinikum
Oberdürrbacher Str. 6 – Haus A4
97080 Würzburg
Germany
uwalter@klin-biochem.uni-wuerzburg.de

Karl Werdan
Department of Medicine III
University Clinics Halle (Saale)
Martin-Luther-University Halle-Wittenberg
Germany

Donald S. Young
Department of Pathology and Laboratory
Medicine
University of Pennsylvania
Philadelphia
Pennsylvania
USA
donaldyo@mail.med.upenn.edu

Brima M. Younis
The Leishmaniasis Research Group/Sudan
The Institute of Endemic Diseases
Medical sciences campus
University of Khartoum
P.O. Box 102
Khartoum
Sudan

1 Plenary Articles

1.1 Proteomics strategies targeting biomarkers for cardiovascular disease

Arjen Scholten and Albert J.R. Heck

Summary

Cardiovascular diseases (CVDs) in the broadest sense refer to the dysfunctional conditions of the heart, arteries and veins that supply oxygen to vital, life-sustaining areas of the body. CVD is one of the main killers throughout the Western world; claiming the lives of more than 17 million people each year. According to the World Health Organization, an estimated 30% of mortalities result from various forms of CVD; whereby a certain proportion can be attributed to genetic defects. For instance, hypertrophic cardiomyopathy often originates from a variety of mutations in genes encoding sarcomere proteins, the structural proteins in heart muscle (myocardium) [1]. However, for other CVDs, such as myocardial infarction, congestive heart failure, atherosclerosis, stroke and deep vein thrombosis, specific gene mutations have not yet been implicated or are only partially responsible for the onset and progression of the disease. Other factors, such as level of stress, food and personal lifestyle can contribute to CVDs. In this light, proteins, the functional and dynamic entities of cells and tissues, are useful indicators to identify crucial mediators of the onset, progression and therapeutic intervention options for the various forms of CVDs. Therefore, proteomics, the global study of protein abundances, function and interaction in an organism at different states is an invaluable technique for gaining insights into CVDs. This burgeoning area of research is aptly termed cardiovascular proteomics [2]. It aims to unravel the molecular basis of cardiovascular function to ultimately improve diagnosis, prognosis and the development of novel therapeutic interventions. Proteome strategies, as used in our laboratory, for identifying CVD-mediated alterations in specific parts of the CVD-proteomes are discussed below.

Protein profiling and targeted proteomic analysis

Nowadays, most proteomics experiments use mass spectrometry (MS) to identify proteins. Identification is achieved by extracting proteins from a given sample and digesting these in peptides using specific proteases, such as trypsin (►Fig. 1.1.1). These peptides are then "sequenced" by tandem MS (MS/MS) by measuring both the intact mass and specific peptide fragmentation masses. Subsequently, both intact and fragment masses are used to identify the protein from which the peptide sequence originated in a database search. To reduce complexity and increase coverage, the sample is separated at either the protein level, peptide level or both prior to MS analysis. For the former, gel-based protein separation is often applied, in one and two dimensions. For the latter, a two-dimensional chromatographic MS/MS (2D-LC-MS/MS) separation is mostly used in which peptides are separated based on parameters such as peptide charge and hydrophobicity.

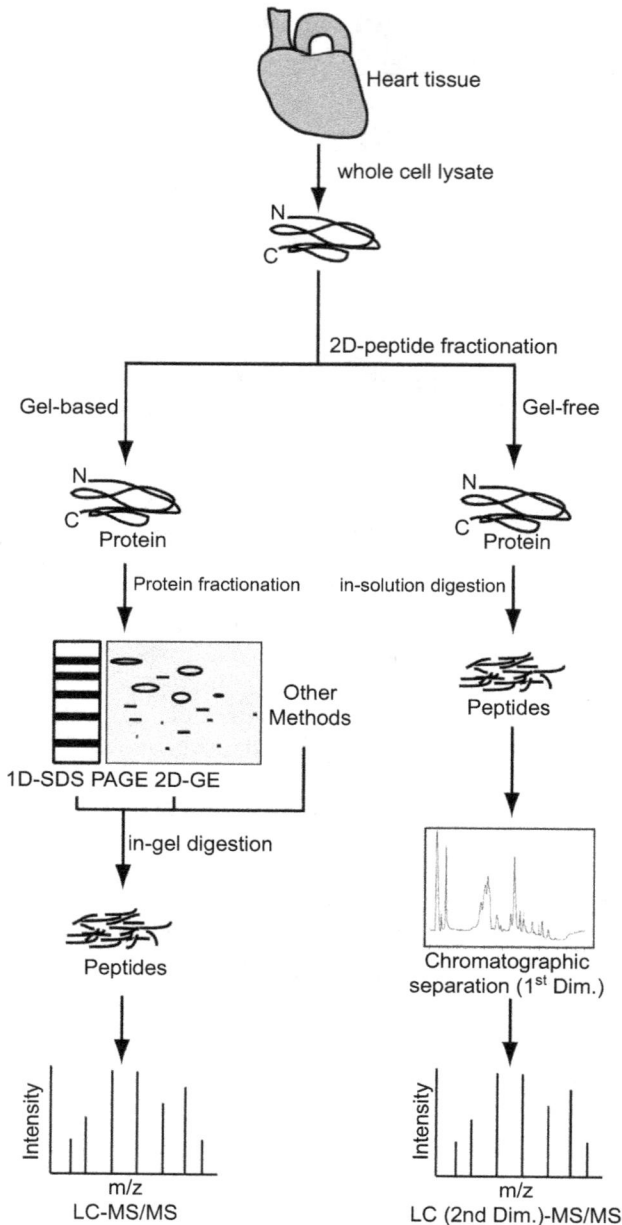

Fig. 1.1.1: General proteomics workflows.
Proteomics can be conducted using gel-based and gel-free approaches. In gel-based approaches, proteins are separated in one or two dimensions prior to digestion into peptides and analysis by MS. Label-free approaches start with digestion of all proteins directly in solution, subsequently using a combination of different orthogonal peptide fractionation steps (e.g. strong cation exchange, reversed phase, hydrophilic interaction, strong anion exchange, etc.) prior to analysis by mass spectrometry.

Protein profiling is the global identification and classification of proteins in a whole organism, a particular organ or a specific cell type. Initially, in clinical proteomics two-dimensional gel electrophoresis approaches were widely used. Over the past decade 2D-LC-MS/MS approaches have gained momentum as they provide more protein identifications. With current instrumentation and infrastructure between 2,000 and 10,000 proteins can be identified corresponding to approximately 20%–90% of the entire expressed proteome. The number of proteins identified is mostly limited by the presence of some highly abundant proteins (e.g. albumin in plasma, myosin in heart tissue), masking the identification of lower abundant proteins. In our efforts to chart the proteome of the human left ventricle we identified well over 3,500 proteins. By using multiple aspects of the MS-generated data, we were able to derive a concentration map of the heart proteome (▶Fig. 1.1.2) [3].

Proteomics aids in biomarker discovery by enabling the search for protein candidates that can serve as specific and sensitive indicators of CVDs. To compare the proteome of two different states, e.g. healthy and diseased, a quantitative aspect is added to the proteomics workflow. This can be done in various ways (for a review see [4]). We mostly use the incorporation of differential stable isotope labels into the proteins or peptides of healthy and diseased samples, which allows the mixing of both states and hence in a 2D-LC-MS/MS platform provides simultaneous identification and quantification of thousands of proteins [5].

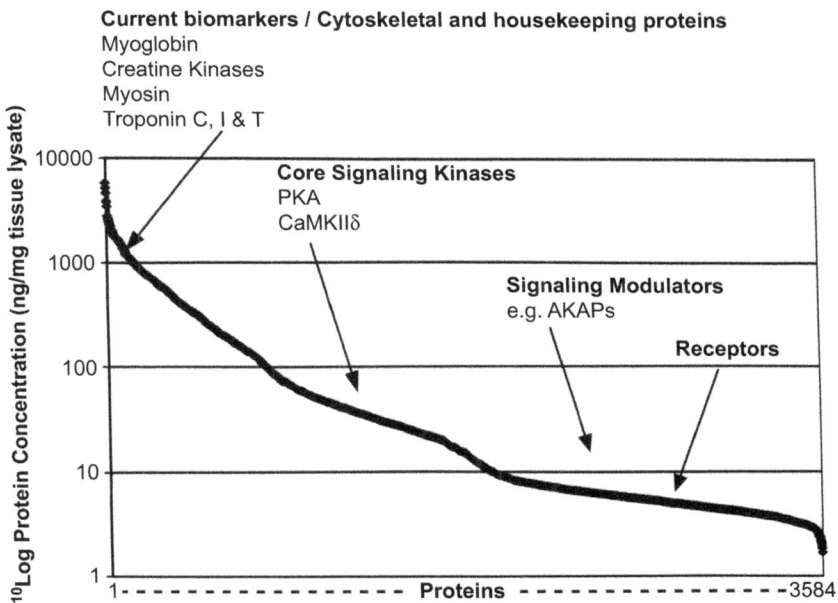

Fig. 1.1.2: A catalogue of protein abundances in the human heart.
Applying complementary proteomics workflows to the analysis of human heart material provided an abundance map of 3,584 proteins. This revealed that in the human heart there is a small subset of highly abundant cytoskeletal and housekeeping proteins. Their abundance is about ~100–1,000-fold higher than crucial signaling kinases. Protein abundance of signaling regulators and receptors was found to be another factor 10–100 lower, making them poorly accessible in standard proteomics experiments.

A number of successful protein diagnostics have emerged from the cardiac proteome. The most definitive of these is targeted at the increased plasma levels of cardiac troponin, which is a primary indicator of myocardial infarction [6]. The success of markers such as cardiac troponin relies on their high abundance in heart tissue, but also on their leakage from the tissue into the plasma. Unfortunately, significant leakage occurs mainly after the myocardium is severely damaged by myocardial infarction, limiting the use of troponin as a biomarker to when serious effects have been initiated. For future diagnostics, targeting so-called "earlier markers" holds great potential to initiate treatment when symptoms are less severe. For this, focus needs to be shifted to proteins that are showing dysfunctional characteristics at that early stage. Potentially these proteins would then also pose as interesting treatment targets. Instead of focusing on the challenging proteomics analysis of plasma, we initially turn towards the organ of interest, i.e. the heart, or search for areas within the blood that have more favorable analysis characteristics for proteomics, such as circulating cells.

One of our goals is to identify the causative molecular determinants of heart failure (HF). In the past, healthy and HF tissue from patients was compared using protein profiling. This global analysis by two-dimensional gel electrophoresis [7] and 2D-LC-MS/MS [8] delivered an enormous candidate list (100+ proteins) for follow-up, making it hard to select proper biomarkers or distinguish the cause/consequence molecular determinants. Therefore we took a hypothesis driven approach, in which we focus on specific aberrant signaling pathways as markers of disease. For HF it is known that β-adrenergic pathways are negatively affected. Going back to our profiling exercise in the human heart [3], we observed that all biomarkers for cardiac disease used today (troponin, creatine kinase, etc.) are in the top 200 proteins of the heart (▶Fig. 1.1.2) and that typical β-adrenergic downstream signaling proteins, such as the important cardiac kinases cAMP-dependent kinase (PKA) and Ca^{2+}/calmodulin dependent kinase reside at already much lower abundance. Crucial signaling modulators of PKA, such as AKAPs, are expressed at even lower levels, which are often below the limit of detection in standard proteomics experiments. Therefore, we developed an enrichment strategy using immobilized cAMP to specifically capture cAMP responsive proteins and their direct protein interactors [9]. This chemical proteomics approach we applied to a discovery set of 5 healthy control left ventricle (LV) tissues and 5 late-stage dilated cardiomyopathic failing LVs. This approach showed specific downregulation of regulatory subunits of PKA, confirming earlier studies [10] (▶Fig. 1.1.3A). Alongside these findings, many other interesting cAMP-mediated events could be monitored using this chemical proteomics approach [11].

In another study, we focus on atherosclerosis. It is well known that a significant percentage of patients that are treated for symptoms of atherosclerosis relapse within 6–12 months. To identify which patients are at risk for relapse, we target accessible molecular determinants in the proteome that can aid in such a relapse risk evaluation, without having to turn to plasma proteomics. Since it is known that the early response to lesions in the vasculature, a risk factor for atherosclerosis, is mediated by circulating cells such as platelets and a variety of lymphocytes, we target individual circulating cell populations directly derived from patients with an ongoing event (e.g. ischemic MI). In comparison with the plasma proteome, the dynamic range of protein abundances of circulating cells are much more compatible with proteomics and can thus yield very specific results (▶Fig. 1.1.3B). The analysis is divided into two phases. We first focus on a small pooled cohort of patients and controls whose circulating cells we analyze using

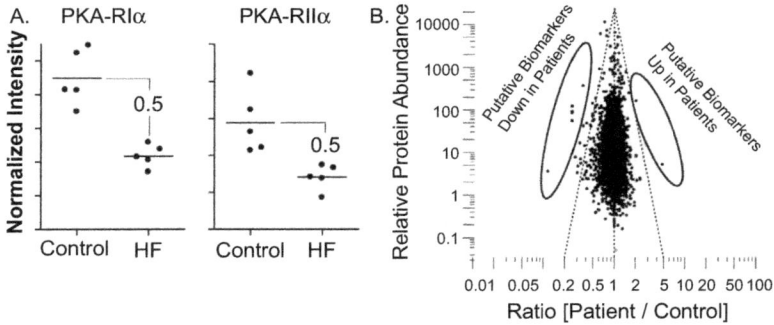

Fig. 1.1.3: Alternative proteomics profiling strategies to target cardiovascular disease-related biomarkers.
A. Using enrichment strategies to analyze differences in the cAMP-binding sub-proteome of heart tissue obtained from healthy people or people having experienced human heart failure revealed a 50% reduction in the capture of cAMP-dependent protein kinase regulatory subunits.
B. Typical large-scale comparison of a circulating cell proteome of atherosclerosis patients and matched controls yields a small subset of altered proteins as putative novel biomarkers.

quantitative protein profiling, allowing us to quantitate well over 3,000 proteins per cell type. This approach averages out inter-patient variability, yielding a more specific candidate list for follow-up in the second phase. At this stage, we use a variety of MS techniques, such as label-free protein profiling and selected reaction monitoring, but also enzyme-linked immunosorbent assay and Western blot assays.

Conclusion

Over the past decade proteomics has entered the area of biomarker discovery, but still needs to overcome some serious bottlenecks. To explore the "holy grail" of early-onset biomarkers, attention initially needs to be shifted to the in-depth analyses of diseased tissue instead of plasma. The challenges of easy accessible detection in patients can be dealt with at a later stage. Hypothesis-driven proteome analysis is one way to gain access to these early markers. Cardiovascular proteomics will prosper in the next 5 years if the field makes optimal use of the rapid technological advancement of proteomic techniques and a strong alliance is built between proteomics experts and clinicians to guarantee the right samples get analyzed in the right way.

References

[1] Maron BJ. Hypertrophic cardiomyopathy centers. Am J Cardiol 2009;104:1158–9.
[2] J. Van Eyk E. Overview: the maturing of proteomics in cardiovascular research. Circ Res 2011;108:490–9.
[3] Aye TT, Scholten A, Taouatas N, Varro A, Van Veen TA, Vos MA, Heck AJ. Proteome-wide protein concentrations in the human heart. Mol Biosyst 2010;6:1917–27.
[4] Bantscheff M, Schirle M, Sweetman G, Rick J, Kuster B. Quantitative mass spectrometry in proteomics: a critical review. Anal Bioanal Chem 2007;389:1017–31.

[5] Boersema PJ, Raijmakers R, Lemeer S, Mohammed S, Heck AJ. Multiplex peptide stable iso-tope dimethyl labeling for quantitative proteomics. Nat Protoc 2009;4:484–94.
[6] Jaffe AS, Ravkilde J, Roberts R, Naslund U, Apple FS, Galvani M, Katus H. It's time for a change to a troponin standard. Circulation 2000;102:216–20.
[7] Corbett JM, Why HJ, Wheeler CH, Richardson PJ, Archard LC, Yacoub MH, Dunn MJ. Cardiac protein abnormalities in dilated cardiomyopathy detected by two-dimensional poly-acrylamide gel electrophoresis. Electrophoresis 1998;19:2031–42.
[8] Kline KG, Frewen B, Bristow MR, Maccoss MJ, Wu CC. High quality catalog of proteotypic peptides from human heart. J Proteome Res 2008;7:5055–61.
[9] Scholten A, Poh MK, van Veen TA, van Breukelen B, Vos MA, Heck AJ. Analysis of the cGMP/cAMP interactome using a chemical proteomics approach in mammalian heart tissue vali-dates sphingosine kinase type 1-interacting protein as a genuine and highly abundant AKAP. J Proteome Res 2006;5:1435–47.
[10] Zakhary DR, Moravec CS, Stewart RW, Bond M. Protein kinase A (PKA)-dependent troponin-I phosphorylation and PKA regulatory subunits are decreased in human dilated cardiomyopathy. Circulation 1999;99:505–10.
[11] Aye TT, Soni S, van Veen TA, van der Heyden MA, et al. Reorganized PKA-AKAP associations in the failing human heart. J Mol Cell Cardiol 2012;52:511–18.

1.2 Influence of pre-examination aspects on result's validity – are ISO 15189 requirements sufficient and clear?

Luděk Šprongl

Summary

The standard EN ISO 15189–2007 is a basic standard of clinical laboratories competence evaluation. In brief, the laboratory's ability to issue valid results is also to be considered within the process of accreditation. The question is whether the above-mentioned standard is sufficient to consider all the processes involved in the creation of results and also whether the fulfillment of all its requirements can be influenced by the laboratory. The standard therefore sufficiently defines the requirements for pre-examination processes but some of the parts of the non-analytical phase cannot be controlled. Therefore by following standard EN ISO 15189–2007, the laboratory does not necessary guarantee the validity of results in cases of tests with more complicated preparation in the collection of more challenging sample collection.

Introduction

Standard EN ISO 15189 was established based on standard EN ISO 17025. It was originally for testing and calibrating laboratories. Considering the specific requirements in healthcare, a specific standard was established in 2002 called EN ISO 15189: Medical laboratories – particular requirements for quality and competence. The standard in the field of clinical laboratories covers not only the analytical process, but also the pre- and post-analytical processes [1,2].

Evaluation of the standard in the pre-analytical phase

What is the pre-analytical phase (pre-examination procedures)?

We can divide this phase in two parts. First, the formulation of a clinical question and the selection of the appropriate examination and second the practical steps of ordering, collection and handling, transportation and reception of samples prior to examination itself [3].

The first phase of the process depends above all on the knowledge and experiences of the consulting physician.

The second part of the pre-analytical phase is mentioned in the standard. It is found in the paragraph 3.11 and Chapter 5.4. The question is whether the requirements are sufficient, clear and evaluable. Does this mean that fulfilling the part of the standard mentioned by the laboratory is sufficient to validate the results and ensure traceability (evaluation of the results in different parts of the country or in another country)? The pre-analytical phase has a crucial influence on the validity of the results. Errors can appear during the preparation of the patient, during the taking of the sample, during transport and during storage.

What is in the standard and where can the problems arise?

Chapter 5.4 has 14 points defining the requirements for the pre-analytical phase in detail. We need to consider what the important points are and how we can comply with them.

5.4.1 – This contains the requirements for the request. The order form is created by the laboratory. The paragraph is clear and easily controllable.

5.4.2 – The requirement for the instruction of the proper collection and handling of primary samples. This is created by the laboratory and is controllable.

5.4.3 – Describes requirements that are necessarily in the primary collection manual. In the first part, the requirements for documentation are mentioned. This part is plain; problems are limited to the intelligibility of the instructions for patients. The second part is devoted to the most important phase – the sample collection, labeling of samples and description of self sample taking. This is again clear and the laboratory can prepare good instructions to fulfill the standard. However the aim of the standard is to ensure valid results and that is endangered in this phase. The third part is more formal – requirements for instruction, completion of a request form, type and amount of the primary sample, timing of collection, transport, sample labeling, and identification of the patient. In practice problems could arise in timing and transport, which could completely devalue the results. The last part is also very important to assure the validity of results – sample storage, requests for additional examinations and repeat examinations. This part of the chapter refers to the laboratory and it is easily controllable [4,5,6].

5.4.5 – Primary sample shall be traceable to an identified individual. In this phase, the problem can only by caused by human error [7].

5.4.6 – This is a very important paragraph about the monitoring of sample transport. There is a need to monitor whether the transport is available at the right time, goes to the right laboratory, and that samples are kept within the specified temperature range, with the right preservatives and in a manner that ensures safety. I consider these requirements, on the basis of my own experience, to be the most problematic part of the pre-analytical phase. The standard is very general at this point and so it is very difficult to keep to it.

Conclusion

It is evident from the above-mentioned points, that almost all aspects that might influence the validity and traceability of results are included in the standard, but are often only as a part of requirements for the documentation.

The significant factors – patient preparation and sample collection – are only obligatory as a part of the primary sample collection manual. However, fulfilling the requirements depends on the responsibility of the physician and where s/he is based (e.g. GP offices). The laboratory can only verify the right use of containers, transport (time, temperature) and agreement between data on the request form and the sample. It is not possible to verify and control patient preparation and the manner of the sample collection.

All of this means that the standard defines requirements for the pre-analytical phase well; however, it is not possible to monitor and control some periods of the non-analytical process. This means that for some examinations with complicated collection, fulfillment of the standard by the laboratory does not guarantee the validity (and traceability) of results.

References

[1] Burnett D. A practical guide to accreditation in laboratory medicine, ACB, London, 2002.
[2] EN ISO 15 189: 2007 Medical laboratories – Particular requirements for quality and competence, ISO, 2007.
[3] Guder WC, Narayanan S, Wisser H, Zawta B. Samples: From the Patient to the Laboratory, 3rd edition, Willey-VCH GmbH Weinheim, 2003.
[4] Astion M. Patient safety focus: mislabelled specimens. Clin Chem News 2010;36:1.
[5] Lippi G, Banfi G, Buttarello M, Cerrioti F, Daves M. Recommendations for detection and management of unsuitable samples in clinical chemistry. Clin Chem Lab Med 2007;45:721–8.
[6] Plebani M. Errors in laboratory medicine and patient safety: the road ahead. Clin Chem Lab Med 2007;45:700–7.
[7] Hernandez J. Patient safety focus: To err is human. Clin Lab News 2010;36:1.

1.3 Posttranslational modifications in tumor diagnosis

Peter Nollau and Christoph Wagener

Summary

Many proteins interact with each other via posttranslational modifications (PTMs). The global status of PTMs reflects the functional state of a cell. In diseases, the cellular functions are disturbed and, as a consequence, the global PTM state of the cell is affected.

The majority of PTMs is bound by special protein domains. These protein domains can be applied in an analytical setting. When protein domains are used on a global scale, the functional state of cells and tissues can be evaluated with high precision.

In a first diagnostic approach, we used recombinant SH2 domains to characterize the global tyrosine phosphorylation state of human malignancies. We were able to group different tumor entities according to their characteristic phosphotyrosine profiles and clinical parameters. In a second approach, the glycosylation state of cells and tissues was characterized by the use of recombinant proteins with carbohydrate recognition domains. We were able to identify a distinct binding pattern in breast cancer extracts, which distinguished node-negative from node-positive primary tumors.

Introduction and results

Posttranslational modifications have a major impact on the cellular function of proteins. The phosphorylation of tyrosine residues mediates central cellular events such as proliferation, motility, apoptosis and senescence. Ubiqutin residues may be attached to proteins as monomers or as ubiquitin chains. Ubiquitination affects a large array of cellular functions such as endocytosis, degradation, and DNA repair. Glycosylation is involved in many cellular functions such as protein folding, signal transduction and cell recognition.

Mammalian genomes encode protein modules, which recognize particular posttranslational modifications. These modules may be used as analytical tools [1]. As posttranslational modifications reflect particular states of cells and tissues, the binding of protein modules to their cognate modifications may allow a precise assessment of normal and disease states. In comparison with antibodies, protein domains recognize functionally relevant structures (▶Fig. 1.3.1).

Here, we present SH2 (SRC homology 2) domains and CRDs (carbohydrate recognition domains) as examples.

Phosphotyrosine kinases mediate the attachment of phosphate groups to tyrosine residues. In some human cancers, tyrosine kinase inhibitors are effective inhibitors of tumor growth. So far, it is not possible to directly measure the activity of tyrosine kinases and phosphatases in human tissue on a comprehensive scale. In order to assess the

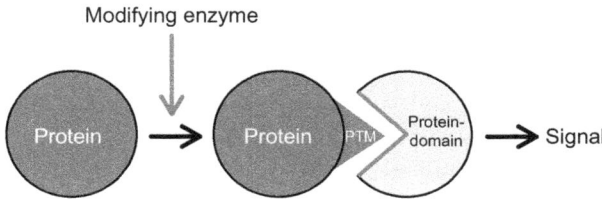

Fig. 1.3.1: Many proteins are enzymatically modified after translation. Posttranslational modifications are recognized by special protein domains, which generate cellular signals.

activity of particular protein kinases and phosphatases, global phosphotyrosine profiling by SH2 domains may be a diagnostically useful alternative [2]. In far Western blots of tumor extracts, labeled SH2 domains reveal distinct binding patterns. An example on breast cancer is presented in ▶Fig. 1.3.2. Taken together our current data suggest that SH2-profiling may serve as a novel and powerful diagnostic tool providing the basis to pin down and interfere with activated, disease-related signaling pathways in human malignancies in the future.

The second example of PTM binding domains concerns the so-called carbohydrate recognition domains of glycoreceptors (CRD) of the C-type lectin family. Structural features of the CRD define the members of the C-type lectin family. The glycoreceptors of the C-type family are expressed by a variety of cells such as macrophages, dendritic cells, NK-cells, and endothelia. Many of the C-type glycoreceptors interact both with infectious agents as well as with cellular glycans.

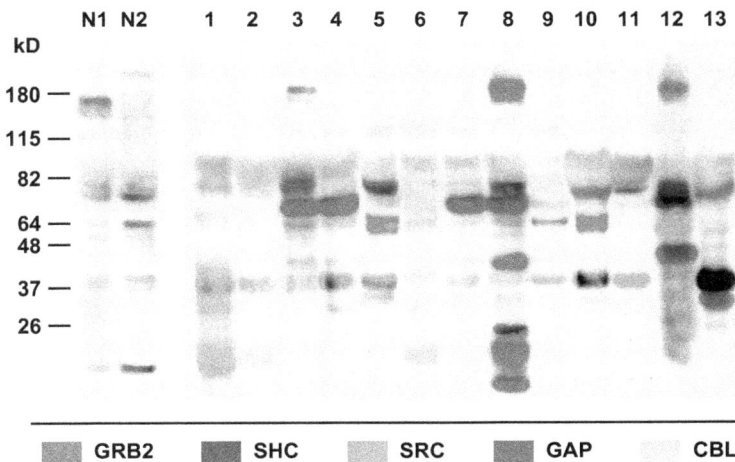

Fig. 1.3.2: Overlay of differential SH2-profiles of human breast cancer specimen (n = 13) in comparison to normal breast tissue (N).
Far Western Blot analyses were performed with the five SH2-domains GRB2, SHC, SRC, RAS-GAP and CBL, respectively. Bound phosphoproteins were detected by chemiluminescence. Applying standard imaging software, signals were transformed to different gray levels and an overlay of the different binding experiments was generated. Compared to normal tissue, the majority of breast cancer specimen shows enhanced and heterogeneous patterns of tyrosine phosphorylation.

In the Consortium of Functional Glycomics, arrays with defined glycostructures have been developed [3]. Using these arrays, the glycan specificity of a number of glycoreceptors has been established. For other glycoreceptors, the specificity has been deduced from the structural analysis of bound compounds. We transfected cells with different fucosyltransferases in order to guide the synthesis of glycans, which bind to glycoreceptors [4]. For many glycoreceptors of the C-type family, the specificity has not yet been established.

Solid tumors induce an inflammatory reaction. Recently, inflammation has been defined as one of the hallmarks of cancer. Among the cells of the inflammatory microenvironment are macrophages, dendritic cells, NK-cells, and endothelia. Each of these cell types expresses glycoreceptors, which potentially interact with the glycans of the tumor cells. These interactions may trigger tumor progression. Characterizing the glycoreceptor-mediated interaction of inflammatory cells of the microenvironment with tumor cell glycans will certainly deepen our knowledge on the molecular mechanisms of tumor progression. If the interactions between the glycoreceptors and the tumor cells are functionally relevant, glycoreceptors may define structural components of the tumor cell membrane, which may be applied as biomarkers either in tissue sections, on sorted cells, or in the circulation. These biomarkers will allow conclusions regarding the biological properties of tumors. In the past, lectins from plants and snails were used to characterize tumor cell glycans. However, according to our own results, the glycan specificities of human glycoreceptors may differ from the specificity of monoclonal antibodies and lectins, though they have been supposed to bind identical structures.

We and others use human recombinant glycoreceptors such as DC-SIGN (dendritic cell specific ICAM-3 binding non-integrin), DC-SIGNR (DC-SIGN-related), SRCL (scavenger receptor with C-type lectin) and MGL (macrophage granulocyte lectin; CD301) to define the glycans and glycoproteins of tumor cells [4,5]. DC-SIGN binds Lewis x structures as well as high mannose residues. In contrast, the homolog DC-SIGNR binds exclusively to high mannose residues. SRCL is a glycoreceptor, which exhibits exclusive specificity for Lewis-related glycan structures. CD301, also termed MGL, binds terminal GalNac residues, which define the tumor associated Tn antigen. Using these and other glycorectors, we identified the distinct binding pattern in breast cancer extracts, which distinguished node-negative from node-positive primary tumors.

Conclusion

PTM binding protein domains can be used in different analytical settings. We established protocols for the staining of formalin-fixed, paraffin-embedded tissues. These protocols work well for some protein domains, but not for each of them. In far Western blots, distinct proteins decorated by defined PTMs can be distinguished. When recombinant glycoreceptors are used for flow cytometry, distinct subpopulations of tumor cells can be defined. It is mandatory to apply appropriate controls to prove binding specificities. Regarding SH2 domains, binding mutants or pretreatment of phosphorylated proteins by phosphatases can be used for negative controls. The binding of glycoreceptors of the C-type lectin family can be performed in the presence and absence of Ca^{++} ions. Alternatively, CRDs can be mutated or deleted.

References

[1] Seet BT, Dikic I, Zhou MM, Pawson T. Reading protein modifications with interaction domains. Nat Rev Mol Cell Biol 2006;7:473–83.

[2] Machida K, Thompson CM, Dierck K, Jablonowski K, Karkkainen S, Liu B, Zhang H, Nash PD, Newman DK, Nollau P, et al. High-throughput phosphotyrosine profiling using SH2 domains. Mol Cell 2007;26:899–915.

[3] Drickamer K, Taylor ME. Glycan arrays for functional glycomics. Genome Biol 2002;3:Reviews 1034.

[4] Bogoevska V, Horst A, Klampe B, Lucka L, Wagener C, Nollau P. CEACAM1, an adhesion molecule of human granulocytes, is fucosylated by fucosyltransferase IX and interacts with DC-SIGN of dendritic cells via Lewis x residues. Glycobiology 2006;16:197–209.

[5] Samsen A, Bogoevska V, Klampe B, Bamberger AM, Lucka L, Horst AK, Nollau P, Wagener C. DC-SIGN and SRCL bind glycans of carcinoembryonic antigen (CEA) and CEA-related cell adhesion molecule 1 (CEACAM1): recombinant human glycan-binding receptors as analytical tools. Eur J Cell Biol 2010;89:87–94.

1.4 Chronic inflammatory disease: a result of complex gene-environment interaction

Harald Renz, Petra Ina Pfefferle and Holger Garn

Summary

Since the end of World War II, a marked increase in the incidence and prevalence of auto-immunities and allergies has been observed. Although genetic factors clearly contribute to the risk of disease development, an even more decisive role is played by the interaction of environmental factors with the host's immune system. In addition to nutritional aspects, the natural microbial environment plays a key role in triggering the development of tolerance. An important role is played by the pre- and post-natal period, during which individuals are highly susceptible to such environment-host interactions. The identification of underlying mechanisms will lead to the exploration of novel bio-markers for the risk assessment of auto-immunities and allergies. Patient stratification based on biomarkers will offer the opportunity to develop preventive strategies for such chronic inflammatory diseases in the future.

The concept of chronic inflammation

Since the end of World War II, there has been a constant increase in almost all chronic inflammatory diseases [1]. The incidence and prevalence of these conditions is increasing, regardless to whether they are of autoimmune or allergic origin. A key component in the immunopathogenesis is the impairment of the development of tolerance. The development of an autoimmune-triggered inflammatory response is prevented by the development of clinical and immunological tolerance. Clinical and immunological tolerance is also a cornerstone in protection against allergic inflammation, which recognizes harmless environmental antigens as dangerous (▶Fig. 1.4.1).

Recent evidence underscores the importance of early life events in the generation of tolerance. It is now well recognized that the underlying mechanisms are the result of active immune responses, utilizing several strategies including the formation of regulatory T-cells, the development of allergic T-cell responses, clonal deletion and others. Further delineation of the kinetics of tolerogenic T-cell functions identified during late pregnancy and early childhood have been determined as important "windows of opportunities" [2]. These data, which were obtained in prospective epidemiological studies as well as in disease-related animal models, highlight the importance of maternal events in regulating and priming early functions in the maturing fetal and neonatal immune system. This raises the question about the contributing maternal and environmental factors that probably have a very decisive role in shaping early immune responses in the offspring. Recent research data now allow a novel insight into this complex regulatory network:

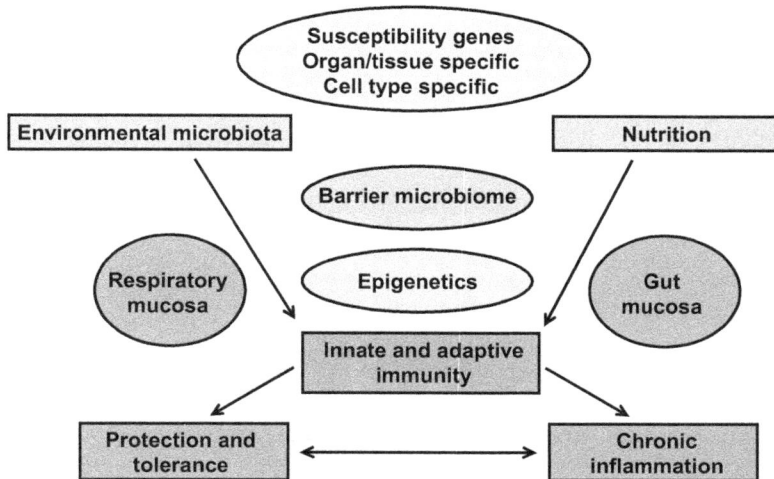

Fig. 1.4.1: The concept of gene–environment interactions to explain the unbalanced host–environment immune interaction in chronic inflammatory disease. Adapted from [9].

The hygiene hypothesis provides the foundation on which to explain the development of tolerance early in life. This concept is based on the epidemiological observation that children born and raised in rural environments are less likely to develop allergic conditions of the respiratory tract during childhood and adolescence [3]. These epidemiological findings were the foundation for further exploration of the biological mechanism of this observation. The model situation for these rural environmental conditions is the (traditional) farming environment in mountain villages of the Alps. Here, a unique qualitative and quantitative exposure to certain microbes has been identified, using state-of-the-art molecular technologies [4,5,6].

These microbes were then further characterized in terms of their immunobiological functions, utilizing suitable animal models of experimental asthma. The microbes of interest include Gram-positive and Gram-negative strains such as *Acinetobacter lwoffii*, *Lactococcus lactis*, *Bacillus licheniformis* and *Lactobacillus GG*. The animal models allowed us to mimic maternal exposure during pregnancy and the effect of bacterial exposures on the programming on the fetal and neonatal immune system [7].

The data obtained in both, epidemiological as well as experimental studies allowed us to draw a number of important conclusions:

(i) The epidemiological data indicate a strong role for the diversity of environmental microbial exposure, as well as in terms of overall bacterial load.

(ii) From experimental animal models the data derived strongly indicate that many different bacterial strains have the capacity to modulate immune responses during early development. However, the detailed impact on clinical phenotype development differs from strain to strain. For example, *Acinetobacter lwoffii* as a Gram-negative bacterium strongly protects airway inflammation, airway hyper-responsiveness and mucus production, whereas other strains impact airway inflammation but not lung function.

(iii) It makes a difference whether exposure occurs in the maternal environment or whether the offspring is directly exposed to the bacteria.

(iv) On the maternal side, initial signaling events originate from the interaction of the microbes with pattern recognition receptors that trigger signaling cascades, resulting in a subclinical proinflammatory immune response that can be systemically detected. This response pattern impacts the biology of placental homeostasis and helps in shaping early immune responses in the offspring.

Early programming of immune responses

How do bacterial antigens program early immune responses? This question is of the upmost importance and has gained recent attention due to the observation that at least certain microbes modulate host-immune gene expression via epigenetic regulation. This has been shown recently by employing the model microbe *Acinetobacter lwoffii*, which has a strong allegro-protective activity. Detailed molecular analysis of the promoter of several Th1 and Th2 relevant immune respondent genes reveals the profound effect on histone acetylating, but not DNA methylation, of the interferon-γ promoter. *In vivo* treatment with a histone deacetylase inhibitor in fact prevented this *Acinetobacter lwoffii*-mediated effect [8].

Conclusion

These data indicate that chronic inflammatory diseases are the result of close and intimate interactions between environmental trigger factors and the host immune system. The fate of this interaction results in either the protection of an inflammatory response with clinical and immunological tolerance development or allows the initiation and progression of chronic inflammation. It will be important to develop biomarkers for early risk assessment that take these environment-host interactions into consideration. With the availability of such biomarkers, the development of novel protective strategies to prevent autoimmunities and allergies will go hand in hand.

References

[1] von Mutius E. Asthma and allergies in rural areas of Europe. Proc Am Thorac Soc 2007; 4:212–6.
[2] Pfefferle PI, Pinkenburg O, Renz H. Fetal epigenetic mechanisms and innate immunity in asthma. Curr Allergy Asthma Rep 2010;10:434–43.
[3] von Mutius E, Radon K. Living on a farm: impact on asthma induction and clinical course. Immunol Allergy Clin North Am 2008;28:631–63x.
[4] Debarry J, Garn H, Hanuszkiewicz A, Dickgreber N, Blümer N, von Mutius E, Bufe A, Gatermann S, Renz H, Holst O, Heine H. Acinetobacter lwoffii and Lactococcus lactis strains isolated from farm cowsheds possess strong allergy-protective properties. J Allergy Clin Immunol 2007;119:1514–21.

[5] Vogel K, Blümer N, Korthals M, Mittelstädt J, Garn H, Ege M, von Mutius E, Gatermann S, Bufe A, Goldmann T, Schwaiger K, Renz H, Brandau S, Bauer J, Heine H, Holst O. Animal shed Bacillus licheniformis spores possess allergy-protective as well as inflammatory properties. J Allergy Clin Immunol 2008;122:307–12.

[6] Ege MJ, Mayer M, Normand AC, Genuneit J, Cookson WO, Braun-Fahrländer C, Heederik D, Piarroux R, von Mutius E; GABRIELA Transregio 22 Study Group. Exposure to environmental microorganisms and childhood asthma. N Engl J Med 2011;364:701–9.

[7] Conrad ML, Ferstl R, Teich R, Brand S, Blümer N, Yildirim AO, Patrascan CC, Hanuszkiewicz A, Akira S, Wagner H, Holst O, von Mutius E, Pfefferle PI, Kirschning CJ, Garn H, Renz H. Maternal TLR signaling is required for prenatal asthma protection by the nonpathogenic microbe Acineto-bacter lwoffii F78. J Exp Med 2009;206:2869–77.

[8] Brand S, Teich R, Dicke T, Harb H, Yildirim AÖ, Tost J, Schneider-Stock R, Waterland RA, Bauer UM, von Mutius E, Garn H, Pfefferle PI, Renz H. Epigenetic regulation in murine offspring as a novel mechanism for transmaternal asthma protection induced by microbes. J Allergy Clin Immunol 2011;128:618–25.

[9] Renz H, von Mutius E, Brandtzaeg P, Cookson WO, Autenrieth IB, Haller D. Gene-environment interactions in chronic inflammatory disease. Nat Immunol 2011;12:273–7.

2   Symposium Articles

2.1 Component-array technology diagnostics: a step forward in the study of the sensitization profile of allergic patients

Juan Manuel Acedo Sanz

Summary

The separation of components diagnosis (component resolved diagnostic testing) is based on the use of purified recombinant allergen or native molecules for the *in vitro* diagnosis of allergy [1]. InmunoCAP ISAC (Phadia) Array technology is the first multiplex *in vitro* diagnostic tool based on allergen components, either specific markers or cross-reactivity markers. The component resolved diagnostic testing identifies individual allergens that cause the disease in each patient, revealing their sensitization profile. Many biological products contain cross-reactive allergens. Panallergen sensitization produces positive results when tested against many allergen extracts. The study of panallergens by array allows us to determine the true cross-reactivity and identify co-sensitized patients [2].

Introduction

The microarray technique applied to the allergy field has generated great expectations because it would theoretically carry out molecular sera analysis of patients for the diagnosis of allergy. Thus it gives clinicians access to detailed information on the profiles of patient sensitization, allowing them to predict the severity and progression of allergic reactions, and provides precise instructions to start immunotherapy.

Based on this technique, InmunoCAP ISAC (Phadia) is the first multiplex *in vitro* diagnostic tool based on the separation of allergenic components. Its technology is based on biochip biotechnology and allows the simultaneous study of 118 allergenic proteins on a single biochip. The physical support used is the microarray and we need a small volume of patient serum to identify specific IgE antibodies against the components of allergens whether natural purified or recombinant. Remember that the array does not determine the allergenic extracts; it is a multiple complex mixture of allergenic proteins and a biological product with major and minor antigenic activity [3].

Component resolved diagnostic testing is a new method for the diagnosis of allergy to help us to identify the molecules responsible of disease. It identifies the allergens that individually cause the disease in each patient, allowing us to see the sensitization profile [4]. In this way it enables us to differentiate the nature of the allergic reaction – i.e. primary or specie-specific – or whether it is a possible cross-reactivity, as there are numerous biological sources that contain cross-reactive allergens.

Allergenic components are classified into protein families according to their function and structure [5]. For this reason, there are species that contain allergenic proteins from specific species that are unique markers of its allergenic source and other proteins with similar structures that are present in biologically-related species. This is the reason why IgE antibodies formed against these structures can bind to the same or similar structure in a protein in several different species causing cross-reactivity.

When this technology is compared to the conventional solid-phase assay for the detection of IgE it has a good correlation, although the sensitivity of some allergens from microarrays may be lower. This may be because the lack of a glucidic part as a consequence of the modification of the bacterial protein production that plays an important role in binding to IgE; or because the panel is incomplete to potential allergens from the source that may cause sensitization [6].

Results

We studied 254 clinically complex patients from the allergy department of the hospital to evaluate the presence of panallergens.

The results were as follows:

- 204 patients had IgE antibodies to panallergens – cross-reactive allergen markers from plants: LTP: 81, PR10: 90, polcalcin: 35, CCD/bromelain: 20 and profilin: 88; cross-reactive allergens markers not from plants: cysteine protease: 24, NPC2: 26, parvalbumin: 47, tropomyosin: 34 and serum albumin: 55.
- 29 patients only had reactivity against specific markers.
- 22 patients showed no reactivity in the analyzer.

The results obtained show that the component separation diagnostic attached to the array technology facilitates the differentiation of panallergens sensitization, identifies co-sensitized patients and improves diagnosis, allowing a better understanding of the sensitization profile. The utility of these results is directed toward a more accurate diagnosis of allergic disease, which with the clinical history will help us to choose and adjust the patient's treatment. To date the results obtained have been comparable with the tests used today, but more results are needed for standardization.

Conclusion

The microarray test for allergen-specific IgE could be considered the method of choice for prospective studies for separating diagnostic components of type I allergy. It could form the basis for the design and monitoring of specific immunotherapy tailored to the patient in the future.

References

[1] Mari A, Alessandri C, Bernardi ML, Ferrara R, Scala E, Zennaro D. Microarrayed allergen molecules for the diagnosis of allergic diseases. Curr Allergy Asthma Rep 2010;10:357–64.
[2] Sastre J. Molecular diagnosis in allergy. Clin Exp Allergy 2010;40:1442–60.

[3] Ferrer M, Sanz ML, Sastre J, Bartra J, del Cuvillo A, Montoro J, Jáuregui I, Dávila I, Mullol J, Valero A. Molecular diagnosis in allergology: application of the microarray technique. J Investig Allergol Clin Immunol 2009;19(Suppl 1):19–24.

[4] Hiller R, Laffer S, Harwanegg C, Huber M, Schmidt WM, Twardosz A, et al. Microarrayed allergen molecules: diagnostic gatekeepers for allergy treatment. FASEB J 2002;16:414–6.

[5] Male D, Brostoff J, Roth D, Roitt I. Inmunología, 7th ed. Madrid: Elsevier-Mosby 2007.

[6] Jahn-Schmid B, Harwanegg C, Hiller R, Bohle B, Ebner C, Scheiner O, Mueller MW. Allergen microarray: comparison of microarray using recombinant allergens with conventional diagnostic methods to detect allergen-specific serum immunoglobulin E. Clin Exp Allergy 2003;33:1443–9.

2.2 Pediatric metabolic syndrome: pathophysiology and molecular mechanisms

Julie Chemin and Khosrow Adeli

Summary

The current obesity epidemic, a condition of excess body fat that may damage health [1], is becoming an increasingly a major health risk worldwide. Over the past few decades, the emergence of this phenomenon in pediatric populations has become a significant challenge for health practitioners. Especially when it is localized within the abdominal region (central obesity), obesity is associated with an increase in metabolic and cardiovascular risk factors and can eventually lead to the development of type 2 diabetes. Obesity is now regarded as one common pathological component of the metabolic syndrome, also called syndrome X or insulin-resistance syndrome. In adults, the International Diabetes Federation [2,3,4] defines the metabolic syndrome as a cluster of risk factors for cardiovascular disease, including visceral adiposity, hypertension, hyperglycemia and dyslipidemia [5]. This definition has been redefined in the pediatric population due to various changes in body size, fat distribution and proportions with age and development. Therefore, children and adolescents are separated into three age groups: age 6 to 9 years, age 10 to 15 years, and age 16 years and older [3]. There are insufficient data available for children below 6 years of age. The aim of the present review is to present the main features of the pediatric metabolic syndrome and to provide some insight into the underlying pathophysiological mechanisms.

Complications associated with pediatric obesity and metabolic syndrome

Pediatric obesity is associated with a number of serious consequences including fatty liver disease, dyslipidemia, hypertension and hepatic insulin resistance.

Fatty liver disease

Liver abnormalities due to obesity are commonly known as nonalcoholic fatty liver disease (NAFLD). Hepatic steatosis, the main feature of NAFLD, is characterized by an excessive amount of intrahepatic triglyceride content. It leads to inflammation with cytokine (interleukin-6 and tumor necrosis factor-alpha) and chemokine production by the adipose tissue and is associated with disorders of glucose (insulin resistance), fatty acid (increased *de novo* lipogenesis) and lipoprotein metabolism (increase in very low-density lipoprotein – triglyceride secretion). Excessive intrahepatic triglyceride content also leads to high expression of several genes involved in metabolic functions, such as hepatic lipase and lipoprotein lipase. However, it is still not fully understood whether NAFLD is a cause or a consequence of metabolic dysfunction [6].

Metabolic dyslipidemia

In obese or overweight children, abnormalities in plasma lipid levels, also referred to as metabolic dyslipidemia, are among some of the early manifestations of metabolic dysfunction and development of the metabolic syndrome. Hypertriglyceridemia, postprandial hyperlipidemia (high chylomicron and very low density lipoprotein [VLDL] remnant levels), hypercholesterolemia (increase in LDL particles), and hypoalphalipoproteinemia (low high-density lipoprotein cholesterol levels) are commonly encountered in overweight and obese children [7,8]. Dyslipidemia likely contributes to further metabolic complications and initiates and/or worsens hepatic and whole-body insulin resistance, significantly increasing the risk of cardiovascular disease later in life.

Hepatic insulin resistance

Insulin resistance occurs when a normal insulin concentration does not produce an adequate insulin response in target tissues, such as adipose tissue, liver and muscle [9]. In response to this abnormality, pancreatic cells produce and secrete more insulin, leading to hyperinsulinemia and a transient normalization of circulating glucose levels, maintaining normoglycemia. Defects in post-receptor signaling in target tissue appear to be the main cause of insulin resistance in obese subjects [10]. Indeed, insulin resistance is associated with reduced activity of the insulin receptor substrates IRS-1 and IRS-2, and lipid/protein kinases (PI3-K and PKB); and increased activity of protein tyrosine phosphatases. Additionally, some inhibitory factors such as suppressors of cytokine signaling (SOCS-1,3) are induced in obese or overweight children, blocking insulin signaling via mechanisms involving the control of IRS-1 phosphorylation and proteasomal degradation of IRS-1 [10].

The central nervous system disorders in obesity

The central nervous system controls food intake and energy homeostasis. The hypothalamic arcuate nucleus, located at the base of hypothalamus, contains the main populations of neurons involved in the regulation of food intake that secrete: appetite-inhibiting pro-opiomelanocortin (POMC), cocaine- and amphetamine-related transcript, appetite-stimulating neuropeptide Y and agouti-related peptide (AgRP). These neurons are sensitive to blood concentrations of adipokines (leptin, adiponectin and others). POMC is a precursor for the anorectic peptide α-melanocyte-stimulating hormone. A lack of POMC in humans leads to obesity. Central injection of AgRP significantly increases food intake. Neuropeptide Y stimulates feeding via some receptor activations and it has effects on other endocrine systems. Another brain region, the dorsal vagal complex, containing the nucleus tractus solitarius and the area postrema, acts as a relay site for short-acting gastrointestinal signals [11].

Conclusion

The increasing occurrence of disorders in children, such as fatty liver disease, dyslipidemia, hypertension and insulin resistance, appear to be a consequence of the obesity epidemic and constitute multiple metabolic risk factors for cardiovascular disease. Much progress has been made in understanding both the causative factors and underlying molecular mechanisms leading to obesity and pediatric metabolic syndrome (►Fig. 2.2.1).

PEDIATRIC METABOLIC SYNDROME

WC > 90th percentile; TG level > $1.7mmol.L^{-1}$; Glucose> $5.6mmol.L^{-1}$; HDL-C<$1.03mmol.L^{-1}$
Blood pressure>130mmHg systolic; Blood pressure> 85mmHgdiastolic

---

NAFLD

Excessive IHTG level
Insulin resistance
High level of VLDL-TG
Inflammatory pathways
activation
Low Adiponectin level

DYSLIPIDEMIA

Hypertriglyceridemia
Hyperlipidemia
Hypercholesterolemia
Hypoalphalipoproteinemia
Hypolipoproteinemia

INSULIN RESISTANCE

Insulin pathway defective
SOCS-1,3 production
PTP1B overproduction
IRSs modified
PI3-K and PKB structural
modification

CNSDISORDERS

Hypothalamus
POMC low level
NPY production
AgRP high level
α-MSH low level

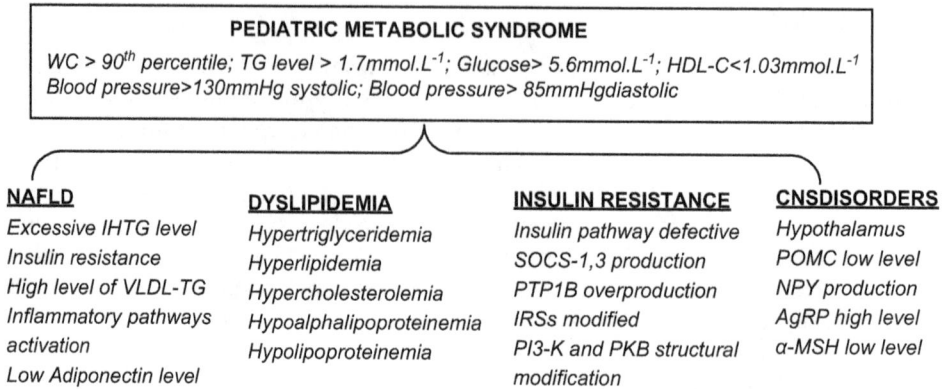

Fig. 2.2.1: Pediatric metabolic syndrome and its clinical complications.
The upper panel lists the key measures used to diagnose metabolic syndrome in children. The onset of the metabolic syndrome in childhood is associated with a number of metabolic and physiological changes. Disturbances occur in many tissues such as adipose tissue, liver and central nervous system. These dysfunctions lead to an increased risk of cardiovascular disease including dyslipidemia, insulin resistance, central adiposity, high blood pressure and nonalcoholic fatty liver disease (NAFLD).

Over-nutrition and reduced physical activity, together with a susceptible genetic background, appear to be the key contributing factors that initiate the development of childhood obesity and metabolic syndrome. Increased abdominal visceral fat leads to enhanced free fatty acid flux and ectopic accumulation of triglyceride in the liver, muscle, and heart. Fatty liver is commonly associated with hepatic insulin resistance, hepatic glucose overproduction, and glucose intolerance. These metabolic defects lead to a prediabetic state in children and adolescents, with a significantly increased risk of developing type 2 diabetes and cardiovascular disease later in adulthood.

Currently, interventions aimed at lifestyle modifications, improved nutrition, and increased physical activity are believed to be the most effective strategies in preventing and treating pediatric metabolic syndrome. Pharmaceutical treatment may be warranted when serious metabolic complications are developed.

References

[1] Lawlor DA, Jago R, Kipping RR. Obesity in children. Part 1: Epidemiology, measurement, risk factors, and screening. BMJ 2008;337:a1824.
[2] The International Diabetes Federation Consensus Worldwide Definition of the Metabolic Syndrome, 2006. www.idf.org/webdata/docs/IDF-Metadef-final.pdf. Accessed 3.17.12.
[3] International Diabetes Federation. Metabolic syndrome in children and adolescents. The IDF consensus definition. IDF, 2007. Available at: http://www.idf.org/webdata/docs/Mets_definition_children.pdf, accessed 3.17.12.
[4] Zimmet P, Alberti G, Kaufman F, Tajima N, Silink M, Arslanian S, Wong G, Bennett P, Shaw J, Caprio S. The metabolic syndrome in children and adolescents: the IDF consensus. Diabetes Voice 2007;52:29–31.

[5] Liu W, Lin R, Liu A, Du L, Chen Q. Prevalence and association between obesity and metabolic syndrome among Chinese elementary school children: a school-based survey. BMC Public Health 2010;10:780.

[6] Klein S, Sullivan S, Fabbrini E. Obesity and nonalcoholic fatty liver disease: Biochemical, metabolic, and clinical implications. Hepatology 2010;51:679–89.

[7] Hersberger M. Dyslipidemias in children and adolescents. Clin Biochem 2011;44:507–8.

[8] Merck Manuals Online Medical Library. Endocrine and metabolic disorders – Lipid disorders. Dyslipidemia, Merck, 2008. Available at: http://www.merckmanuals.com/professional/sec12/ch159/ch159b.html, accessed 3.17.12.

[9] Adeli K, Theriault A, Kohen-Avramoglu R. Emergence of the metabolic syndrome in childhood: an epidemiological overview and mechanistic link to dyslipidemia. Clin Biochem 2003;36:413–20.

[10] Adeli K, Meshkani R. Hepatic insulin resistance, metabolic syndrome and cardiovascular disease. Clin Biochem 2009;42:1331–46.

[11] Bloom SR, Murphy KG. Gut hormones in the control of appetite. Exp Physiol 2004;89:507–16.

2.3 Urinary 8-hydroxydeoxyguanosine as a biomarker of microangiopathic complications in type 2 diabetic patients

Amr Fattouh, Nermine H. Mahmoud, Shereen H. Atef and Abd El Rahman Gaber

Summary

Background: Reactive-oxygen species (ROS) produced either endogenously or exogenously can attack lipid, protein and nucleic acid simultaneously in living cells. Increased oxidative stress induced by hyperglycemia may contribute to the pathogenesis of diabetic complications. Urinary 8-hydroxydeoxyguanosine (8-OHdG) has been reported to serve as a sensitive biomarker of oxidative DNA damage.

Objective: To evaluate urinary 8-OHdG as a marker for diabetic microangiopathic complications and to correlate its levels with the severity of diabetic nephropathy and retinopathy.

Subjects and methods: The study included 50 patients with type 2 diabetes mellitus and 30 non-diabetic age- and sex-matched control individuals. 8-OHdG, urine creatinine and urinary albumin excretion (UAE) rate were measured in all patients and control subjects. Both 8-OHdG and UAE rate were assayed by immunoassays. Assessment of glycemic control in patients was achieved by measurement of HbA_{1c}. All of the patients underwent direct opthalmoscopy and photography with pupils dilated.

Results: There was a highly significant difference between different groups of type 2 diabetic patients classified according to retinopathy, and controls as regards 8-OHdG ($p < 0.01$), and albumin to creatinine ratio ($p < 0.01$). Statistical comparison between groups of patients classified according to albumin to creatinine ratio using analysis of variance test revealed a highly significant difference regarding 8-OHdG, ($p < 0.01$). Using a ROC curve, the diagnostic utility of urinary 8-OHdG in discrimination of diabetic patients with retinopathy from those without retinopathy at a cut-off level of 34.4 µg/g creatinine had a diagnostic sensitivity of 92.9%, specificity of 86.4% and efficacy of 90%.

Conclusion: Measuring 8-OHdG is a novel and convenient method for evaluating oxidative DNA damage. Diabetic patients, especially those with advanced nephropathy and retinopathy, had significantly higher levels which may contribute to the development of the microvascular complications of diabetes.

Introduction

Vascular complications are the leading cause of morbidity and mortality in patients with diabetes. Considerable evidence has been accumulated to suggest that the production of reactive oxygen species (ROS) and lipid peroxidation are increased in diabetic patients, especially in those with poor glycemic control [1]. Oxidative stress may be crucial for the development of diabetic microvascular complications [2].

Intracellular ROS can cause strand breaks in DNA and base modifications, including the oxidation of guanine residues to 8-hydroxydeoxyguanosine (8-OHdG), so 8-OHdG may serve as a sensitive biomarker of intracellular oxidative stress *in vivo* [2].

The urinary level of this molecule is now considered a biomarker of the total systemic oxidative stress *in vivo*. Urinary excretion of 8-OHdG has been shown to be higher in both type 1 and type 2 diabetic patients compared with non-diabetic individuals [3].

Recently, the content of 8-OHdG in urine and that of the isolated mononuclear cells from type 2 diabetic patients with either retinopathy or nephropathy were reported to be much higher than those in patients without these complications [4].

The aim of the present study was to investigate the levels of urinary 8-OHdG in type 2 diabetic patients and to correlate its levels with the severity of diabetic nephropathy and retinopathy in an attempt to determine the possible contribution of oxidative DNA damage to the pathogenesis of diabetic microangiopathic complications.

Patients and methods

Patient group

This study was conducted on 50 type 2 diabetic patients. They were selected from the outpatient clinic of the ophthalmology department of Ain Shams University Hospital. All of the patients fulfilled the diagnostic criteria for type 2 diabetes [5]. None of the patients were cigarette smokers or suffered from cancer.

Patients were on oral hypoglycemic agents. All patients had undergone fundoscopic examination using an ophthalmoscope, and were classified into 3 groups according to albumin/creatinine ratio in urine:

(i) patients with normoalbuminuria: albumin/creatinine up to 30 µg/g creatinine;
(ii) patients with microalbuminuria: albumin/creatinine ratio ranges from 30–300 µg/g creatinine; and
(iii) patients with macroalbuminuria, albumin/creatinine ratio is >300 µg/g creatinine.

Patients were also classified into 3 groups according to retinopathy, using the Davis classification [6]:

(i) diabetics with no retinopathy;
(iii) diabetics with non-proliferative retinopathy; and
(iii) diabetics with proliferative retinopathy

Control group

Thirty age-matched apparently healthy subjects were included in the study. All had normal fasting blood glucose levels. All individuals studied were subjected to the following:

- full medical history; and
- laboratory investigations including: urinary 8-OHdG, creatinine, albumin and glycated hemoglobin (HbA$_{1c}$).

Analytical methods

Urinary creatinine was determined using the Synchron CX-9 Autoanalyzer (Beckman Inst. Inc., CA, USA), after dilution of urine samples. Urinary albumin was determined using a Randox immunoturbidimetric kit (Randox Laboratories Ltd., Ardmore, Co. Antrim, UK). HbA$_{1c}$ was determined in the patient group using the column chromatography technique, a product produced by Tecodiagnostics (Anaheim, CA, USA). Urinary 8-OHdG was determined using a competitive enzyme-linked immunosorbent assay kit (S-OhdG Check; Japan Institute for the Control of Aging, Shizuoka, Japan).

Statistical analysis was carried out using the SPSS software package (Echo soft Corp., USA).

Results

Statistical comparison between the control group and groups of patients classified according to retinopathy showed a highly significant difference regarding 8-OHdG and albumin/creatinine ratio (F = 5.6 and F = 5.2, respectively; $p < 0.01$). Statistical comparison between patients with macroalbuminuria and those with microalbuminuria showed a significant difference with regards to 8-OHdG level ($p < 0.05$) (▶Tab. 2.3.1). There was also a significant difference in urinary 8-OHdG between patients with macroalbuminuria and patients with microalbuminuria as compared to patients with normoalbuminuria ($p < 0.0l$ and $p < 0.05$, respectively) (▶Tab. 2.3.1).

When the relationship between severity of diabetic retinopathy and U8-OHdG was assessed, its levels were significantly higher in patients with non-proliferative retinopathy (t = 2.8, $p < 0.05$) and those with proliferative retinopathy (t = 3.2, $p < 0.01$), (▶Tab. 2.3.1).

Tab. 2.3.1: Statistical comparison between the different studied groups regarding 8-OHdG using Student's t-test.

Compared Groups	8-OHdG	(µg/g creat)
	T	p
Micro-vs. normo albuminuria	2.0	<0.05
Macro-vs. normo albuminuria	3.4	<0.01
Macro-vs. micro albuminuria	2.5	<0.05
Non-proliferative vs. no retinopathy	2.8	<0.05
Proliferative vs. no retinopathy	3.2	<0.01
Non-proliferative vs. proliferative retinopathy	1.5	>0.05

$p < 0.01$ = highly significant difference, $p < 0.05$ = significant difference, $p > 0.05$ = non-significant difference.

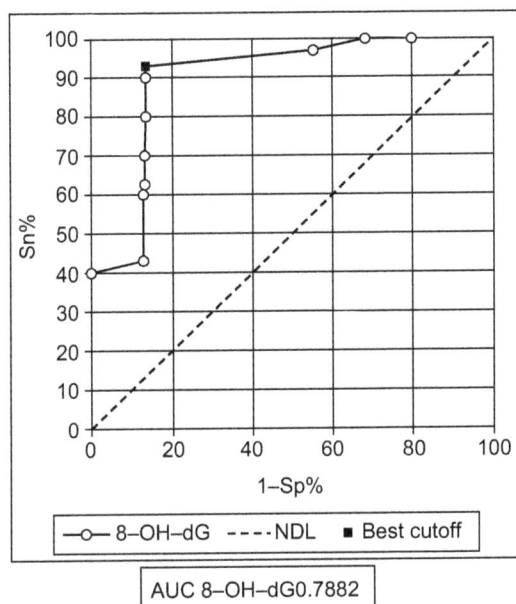

Fig. 2.3.1: ROC curve analysis showing the diagnostic performance of 8-OHdG for discriminating patients with retinopathy from those without retinopathy.

ROC curve analysis was done to assess the diagnostic utility of urinary 8-OHdG in diabetic patients with retinopathy vs. patients without retinopathy (▶Fig. 2.3.1). At a cut-off level of 34.4 µg/g creatinine, the diagnostic sensitivity was 92.9%, the specificity was 86.4% and the efficacy was 90%, with area under the curve of 0.7882.

Conclusion

Hyperglycemia causes oxidative damage to DNA, which might play a role in the pathogenesis of diabetic complications. Increased 8-OHdG (a product of oxidative stress) levels in type 2 diabetic patients have been recorded. Higher levels are found in patients with advanced complications. Accordingly, 8-OHdG is considered a reliable marker for evaluating oxidative DNA damage and can be used as a biomarker of microvascular complications in patients with type 2 diabetes.

References

[1] Leinonen J, Lehtimäki T, Toyokuni S, Okada K, Tanaka T, Hiai H, Ochi H, Laippala P, Rantalaiho V, Wirta O, Pasternack A, Alho H. New biomarker evidence of oxidative DNA damage in patients with non-insulin-dependent diabetes mellitus. FEBS Lett 1997;417:150–2.
[2] Wu LL, Chiou CC, Chang PY, Wu JT. Urinary 8-OhdG: a marker of oxidative stress to DNA and a risk factor for cancer, atherosclerosis and diabetics. Clin Chim Acta 2004;339:1–9.

[3] Hinokio Y, Suzuki S, Hirai M, Suzuki C, Suzuki M, Toyota T. Urinary excretion of E-oxo-7, 8-dihydro-2'-deoxyguanosine as a predictor of the development of diabetic nephropathy. Diabetologia 2002;45:877–82.

[4] Nishikawa T, Takayuki S, Shinsuke K, Kazuhiro S, Takahumi S. Evaluation of Urinary 8-hydroxydeoxyguanosine as a novel biomarker of macrovascular complications in type 2 diabetes. Diabetes Care 2003;26:1507–12.

[5] American Diabetes Association. Diagnosis and classification of diabetes mellitus. Diabetes Care 2006;29(1):43–8.

[6] Davis MD, Kern TS, Rand LI. Diabetic retinopathy. In: Albertie KGMM, Zimmet P, Defranzo RA (eds). International textbook of diabetes mellitus. 2nd ed., Chichester, UK: John Willey, 1997; pp. 1413–46.

2.4 Biological variation data: the need for appraisal of the evidence base

William A. Bartlett

Summary

Biological variation data have many established applications, ranging from the setting of quality standards for analyses in clinical laboratories to the assessment of significance of change in serial results (reference change values). Given the importance of these applications, there is an imperative that these fundamental data are fit for purpose. If the data are flawed in any way, or inapplicable to the population to which a measurement is being applied, then the application must be considered to be erroneous. It can be argued that these data are in fact reference data and as such require standards for their production, reporting and transmission to ensure that they are fit for purpose and applied appropriately.

Introduction and outcome

Review of the literature on biological variation identifies a significant volume of work stretching back some 40 years. Application of these data must follow an objective and critical assessment of the data on a case-by-case basis. Ideally the data sets need to be accompanied by adequate descriptions of the populations studied, derived using an appropriate experimental design, accompanied by details of analytical methods used in terms of analytical performance and specificity, alongside evidence of application of appropriate statistical methods to the component data sets. There are parallels here to the requirements for the production and reporting of reference values as identified by the International Federation of Clinical Chemistry. Users should not underestimate the complexity of biological variation data. To enable an appropriate evaluation of biological variation data, a critical appraisal checklist is required to enable veracity of existing and newly-published biological variation data in the growing evidence base.

The variance observed in clinical laboratory measurements (CV_T) arise from the sum of the analytical variability (CV_A), the within subject variability (CV_I) and the between subject variability (CV_G). Using appropriately designed studies, it is possible to dissect these components of variance to specifically enable identification of the non-analytical variances:

$$CV_T = CV_A + CV_I + CV_G$$

There are well-defined approaches to the generation of these data and their application [1]. The data are used not only to establish quality standards for laboratory measurements but in many other ways, ranging from derivation of reference change values to assess the significance of change between consecutive measurements, to the assessment of the utility of reference intervals. These fundamental data therefore have many applications that underpin the practice of clinical laboratories. It can be argued that these data are important reference data that should always be accompanied by metadata describing the characteristics of the population studied, method of analysis and other relevant data to enable commutability across geography and time.

Users of published data need to be concerned about the validity and robustness of data, whether they can be applied to their practice, the method of analysis, time frame of the experiment and the implications of error in the estimates of component variances. A minimum data set should therefore accompany the estimates of variation. The methods for the generation of the data should be fit for purpose and reported in sufficient detail to enable their application. Interesting parallels arise with the requirements for reference values as outlined theory of reference values [2]. This highlights the requirement for standards for the production, reporting and transmission of biological variation data, with a further requirement to derive a standard data "archetype" to enable the commutability and application of the data.

There is literature relevant to the derivation and application of biological variation stretching back over 40 years that continues to grow in volume. The utility of the historical and current literature will depend upon the design of the experiment set up to derive the indices, the veracity of the data analysis in terms of outlier exclusion and determination of the homogeneity of the variance, and not least the assay characteristics. Specificity of methods is an important determinant of biological variation data. Generational changes in assay technology applied to large molecules, for instance, results in more specific assays that are less affected by renal clearance of fragments (e.g. parathyroid) and might be expected to show differing degrees of variation in chronic kidney disease patients.

Biological variation data have been collated from the literature in databases [3] and reviews [4]. While useful resources, they do not necessarily have the granularity required to enable a critical evaluation of the quality of the data. Referral back to the original publications is essential, with critical appraisal checklist that can also be used to set the standard for new publications. A further degree of complexity is the variability in data between diseased and non-diseased subjects. A systematic review of biological variation data pertaining to glycated hemoglobins [5] highlights many of the issues raised here.

Conclusion

Biological variation data are in fact reference data that have important applications in clinical laboratory settings that demand standards for production, reporting and transmission. Development of a critical appraisal checklist will enable users to assess the existing literature and reviewers to deliver a standard for future publication (www.biologicalvariation.com). The delivery of a standard archetype is required that enables the transmission of relevant metadata with the biological variation indices. Users of data presented in compiled databases and reviews need to be aware of the complexity of the data and be critical in its application.

References

[1] Fraser CG, Harris EK. Generation and application of data on biological variation in clinical chemistry. Crit Rev Clin Lab Sci 1989;27:409–37.
[2] Solberg HE. Approved recommendation (1986) on the theory of reference values. Part 1. The concept of reference values. J Clin Chem Clin Biochem 1987;25:337–42.
[3] Westgard. www.westgard.com/biodatabase1.htm. Westgard QC, Madison, Wisconsin, 2012.
[4] Ricós C, Iglesias N, García-Lario JV, Simón M, Cava F, Hernández A, Perich C, Minchinela J, Alvarez V, Doménech MV, Jiménez CV, Biosca C, Tena R. Within subject biological variation in disease: collated data and clinical consequences. Ann Clin Biochem 2007;44:343–52.
[5] Braga F, Dolci A, Mosca A, Panteghini M. Biological variability of glycated hemoglobin. Clinica Chimica Acta 2010;411:1606–10.

2.5 *Recovery*ELISA – a newly-developed immunoassay for measurement of therapeutic antibodies and the target antigen during antibody therapy

Pavel Strohner, Dieter Sarrach, Jens G. Reich, Antonia Staatz, Astrid Schäfer, Jens-Oliver Steiß, Thomas Häupl and Gunther Becher

Therapeutic antibodies (TAbs) are increasingly being used in clinical use and have a huge financial impact on health systems worldwide. Many clinical studies on TAbs against a human antigen (target protein) are listed and will effect treatments against various diseases, e.g. autoimmune and oncological diseases. During diagnostics, TAb treatment is always a challenge and in many cases the target protein cannot be measured reliably with conventional tests. Therefore, a solution to this problem is the newly-developed *recovery*ELISA, which is an immunoassay for measuring TAb and the target antigen during antibody therapy. This test accurately handles the disturbing influence of the TAb in patient samples. The *recovery*ELISA is a universal assay technology for monitoring therapy with TAb and delivers three results from just one test:

- free levels of antigen (if present in serum);
- therapeutic antibody levels; and
- a general dose-response interaction.

The so-called "recovery curve" is the final result of *recovery*ELISA because it shows the relationship between therapeutic antibody level and the neutralization rate of target protein. With the aid of two independent calibrations, an evaluation procedure is carried out to determine the concentration of antigen and TAb in samples. This evaluation is performed by non-linear regression using the Michaelis-Menten model for the standard curve and the Langmuir isotherm for the TAb curve.

The *recovery*ELISA is a newly-developed immunoassay technology for monitoring antibody therapies in human serum. It combines a Sandwich-ELISA (solid-phase immunoenzymetric assay) for the measurement of antigens (target proteins) with a competitive enzyme-linked immunosorbent assay (ELISA) for the detection of therapeutic antibodies [1,2].

This means a two-dimensional calibration is performed:

(i) Antigen calibrators without and with additional TAb against optical density; and
(ii) TAb calibrators against antigen recovery.

The presence of therapeutic antibodies in serum samples considerably influences conventional immunoassays [3]. The recovery of the antigen is systematically decreased, depending on the concentration of TAb calculated in relation to a calibration curve without TAb.

Recalculated amounts of antigen in standard curves and the spiked amount of antigen in samples (with and without the addition of TAb) result in straight lines (▶Fig. 2.5.1). The slope of this line is the recovery of antigen (▶Fig. 2.5.2) and enables the calculation of free antigen under the influence of the therapeutic antibody in the sample. Since the slope of the line is uniform, the two-dimensional calibration can be transferred to a one-dimensional calculation.

The recovery curve (▶Fig. 2.5.2) itself shows the rate of free antigen, depending on TAb addition. The neutralization rate of target protein is complementary to the recovery. The higher the TAb addition, the lower the recovery and the higher the neutralization rate of the antigen (there is a dose-response interaction).

In unknown samples, the recovery can be estimated from the addition of a certain amount of the antigen [4,5]. The *recovery*ELISA method can be adapted for almost all therapeutic antibodies.

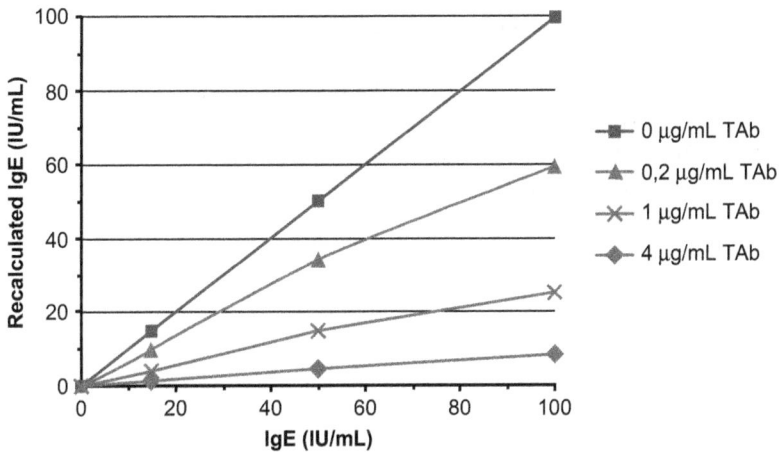

Fig. 2.5.1: Recovery of IgE (Antigen) in dependence of TAb addition.

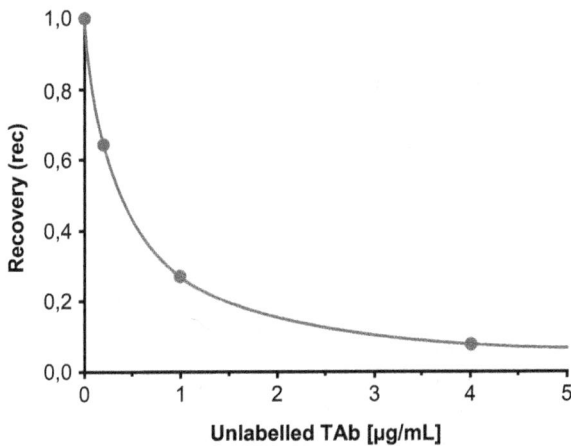

Fig. 2.5.2: Recovery curve: rec = $1/(1 + k1*q1)$ + NSB; rec: Recovery; k1: equilibrium constant of reaction of TAb with antigen; q1: total TAb; NSB: non-specific binding.

For the development of *recovery*ELISAs, the following preconditions are necessary:

1. Sandwich ELISA must be available;
2. general assumptions:
     (i) there is excess of TAb in relation to target protein,
     (ii) this is excess complex TAb and target protein in relation to the antibodies captured on the microplate, and
     (iii) there is equilibrium.

The application of the *recovery*ELISA is demonstrated on clinical serum samples of patients treated with omalizumab (an anti-IgE) and for adalimumab (an anti-tumor necrosis factor-alpha drug).

An assay procedure example of *recovery*ELISA for IgE/omalizumab (TAb)

The test takes place in microplate cavities that are pre-coated with capture antibody. IgE standard (with and without TAb) or the sample (restocked with IgE) as well as the signal antibody – *horseradish peroxidase (HRP)*-conjugate are dispensed into the cavities. During incubation, the capture antibody and the signal antibody – peroxidase-conjugate react with two different binding epitopes on IgE. Thus an immobilized complex of capture antibody, IgE and signal antibody – HRP-conjugate (sandwich) is formed. Additionally, there is a competitive reaction between the unlabeled TAb that is already in the samples and the IgE-standards [6]. The incubation is discontinued after 16 hours by washing followed by enzyme reaction. The optical density rises with increasing concentration of IgE and falls with the rising concentration of unlabeled omalizumab.

Conclusion

The *recovery*ELISA enables the monitoring of therapeutic antibodies and the target antigen during treatment. Irregular recovery also indicates auto-antibodies. The *recovery*ELISA is a general assay concept that can be adopted for every therapeutic antibody or its derivatives, whether the target protein is soluble in serum or not.

To summarize, when a standard curve for the antigen and a recovery curve for TAb is applied, the *recovery*ELISA can estimate the concentration of TAb as well as free antigen in a patient's serum. Therefore *recovery*ELISA is expected to provide the fundamental information necessary for the therapeutic application of those antibodies.

References

[1] Strohner P. Patent: EP 06828568.3, Immunoassay for the simultaneous immunochemical determination of an analyte (antigen) and a treatment antibody targeting the analyte in samples (Recovery immunoassay), filed November 29, 2006.
[2] Steiß JO, Strohner P, Zimmer KP, Lindemann H. Reduction of the total IgE level by omalizumab in children and adolescents. J Asthma 2008;45:233–6.
[3] Hamilton RG, Marcotte GV, Saini SS. Immunological methods for quantifying free and total serum IgE levels in allergy patients receiving omalizumab (Xolair) therapy. J Immunol Methods 2005;303:81–91.

[4] Feldman HA. Mathematical theory of complex ligand binding systems at equilibrium: some methods for parameter fitting. Anal Biochem 1972;48:317–38.

[5] Reich JG. Curve fitting and modeling for scientists and engineers. New York: Mc Graw-Hill; 1991.

[6] Strohner P, Sarrach D, Reich JG. Use of a modified binding model for the investigation of affinity dependence on antibody concentration in immunoassay systems. J Immunol Methods 1997;203:113–22.

2.6 Genotypic prediction of HIV-1 tropism from plasma and peripheral blood mononuclear cells in the clinical routine laboratory

Christian Paar, Maria Geit, Herbert Stekel and Jörg Berg

Summary

HIV-1 entry is initiated by the binding of viral particles to CD4 and co-receptor molecules, mainly CCR5 (R5) and CXCR4 (X4) on the host cell membrane. Maraviroc, an R5 antagonist that blocks HIV-1 entry, has been introduced into clinical practice. As Maraviroc treatment is restricted to patients harboring viruses using R5 as co-receptor, tropism testing is required prior to therapy. We report on the development of a sequencing assay for the genotypic HIV-1 tropism determination, which allows both HIV-RNA from plasma and HIV-DNA from peripheral blood mononuclear cells (PBMC) to be examined. The online bioinformatics tool geno2pheno$_{[coreceptor]}$ was used for tropism prediction. The assay was validated by the examination of reference samples representing HIV-1 group M subtypes A–H. This yielded up to 100% matching genotypic tropism prediction compared to the respective phenotypes of the reference samples. Testing for accuracy and assaying of 188 plasma samples and 331 PBMC samples revealed robustness and reliability across many HIV-1 group M subtypes, including some circulating recombinant forms. Assaying of 150 PBMC samples with undetectable viral loads revealed that those samples can also readily be examined. Pair-wise examinations of 103 plasma and PBMC samples from identical collection tubes showed a concordance of viral and proviral tropism determination of 90.3%. With the limitations of population-based sequencing and of using a bioinformatics tool for tropism prediction, the assay was found to be suitable for genotypic HIV-1 tropism determination in a routine clinical laboratory.

Introduction

HIV enters target cells by using CD4 molecules and the chemokine receptors CCR5 (R5) and CXCR4 (X4) as major co-receptors. Depending on co-receptor usage, HIV tropism is classified as R5, X4 or R5X4 (dual/mixed) [1]. Maraviroc blocks HIV entry by binding to CCR5 molecules [2]. In clinical trials, Maraviroc has been shown to reduce viral loads and was introduced into clinical practice as an oral HIV-entry inhibitor. Administration of Maraviroc is limited to patients exclusively harboring HIV R5 strains; therefore, tropism testing is mandatory in clinical laboratories.

We developed a sequencing assay for the genotypic determination of HIV tropism for clinical routine laboratory testing. The assay was configured to allow examination of HIV-RNA from plasma and HIV-DNA from peripheral blood mononuclear cells

(PBMCs) for patients with undetectable viral loads, as Maraviroc may offer additional favorable therapeutic options in these patients.

Materials and methods

Plasma and PBMC samples from the Austrian HIV cohort and the AREVIR HIV-1 group-M subtype reference panel were used throughout the study.

RNA and DNA were extracted using standard methods. Viral loads were assessed with the COBAS TaqMan 48 HIV-1 test (Roche, Austria). RNA was reverse transcribed and amplified by touch-down nested polymerase chain reaction (Onestep RT-PCR kit, Qiagen, Germany) using novel primers that had been designed: V3-forward 5'-GTCAGCACAGTACAATGYACACATGG-3'; and V3-reverse 5'-AGTGCTTCCTGCTGCTCCYAAGAACCC-3'. DNA was amplified by nested primer polymerase chain reaction with the primers listed above. The primers V3-nested-for 5'-TGTTAAATGGCAGYCTAGCAG-3' and V3-nested-rev 5'-TGGGAGGGGCATACA TTG-3' were used in the second polymerase chain reaction and the sequencing.

Sequencing was carried out with BigDye® (Applied Biosciences, Foster City, CA, USA) termination followed by sequence analysis on the ABI 3130 instrument (Life Technologies Corp., Germany). Sequences were submitted to the bioinformatics tool geno2pheno$_{[coreceptor]}$ for HIV-tropism prediction [3]. A cut-off at false positive rate (FPR) of 5.75% was used for the examination of the reference panel and the comparison of paired plasma and PBMC samples [4]. Clinical samples were analyzed with the German recommendations interpretation scheme [5]. X4 tropism is suggested at FPR ≤ 5.75%, ambiguous tropism at FPR between 5.75% and 20%, and R5 tropism at FPR >20%, respectively.

Results and discussion

Examination of the HIV-1 subtype reference panel returned 100% matching results in the 11 samples with available phenotypic data. Thus, the assay showed unbiased sequencing and tropism determination across HIV-1 group M subtypes A–H.

Sequencing accuracy was assessed by assaying one RNA sample (1000 copies/mL) 8 times, independently. In 6 out of 8 runs, identical sequences were obtained. One run showed one wobble base and one run yielded three wobble bases. The wobble bases did not affect the determination of tropism.

The assay was clinically evaluated by testing 188 plasma and 331 PBMC samples, 150 of which did not exhibit any viral load. A wide spectrum of HIV group M subtypes was covered with those clinical samples: A–D, F, G, and CRF01 and CRF02. Subtype B was the most frequent. All samples returned interpretable sequencing results to allow the prediction of tropism (▶Fig. 2.6.1). R5 tropism was more prevalent in plasma samples than in PBMC samples, which is in line with previous observations [6].

To specifically compare tropism determination from PBMC HIV-DNA and plasma HIV-RNA, 103 paired samples from identical collection tubes were assessed. This revealed a concordance of 90.3% (▶Tab. 2.6.1). Our data are perfectly in line with a recent study showing a 90% concordance for tropism determinations in the plasma and

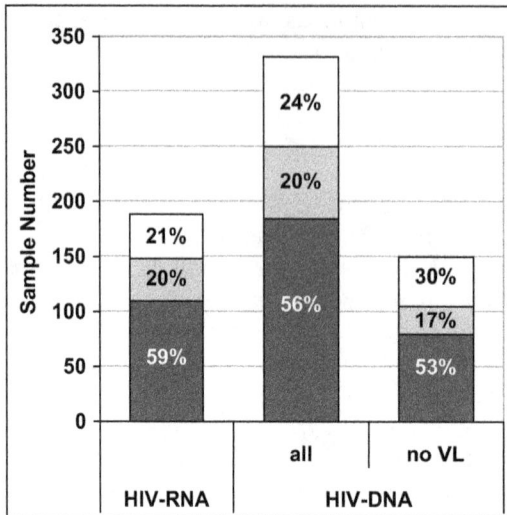

Fig. 2.6.1: Genotypic tropism testing of clinical samples.
HIV-RNA from plasma (viral load 1.4×10^2–7×10^7 copies/mL) and HIV-DNA from PBMC (no viral load–2×10^6 copies/mL) were sequenced and tropism was determined with g2p as described in the text. The clinical interpretation scheme "German recommendations" was applied. White filling, X4 tropism; grey filling, ambiguous tropism; black filling, R5 tropism.

Tab. 2.6.1: Tropism determination of paired plasma HIV-RNA and PBMC HIV-DNA samples[a] (n = 103).

		Plasma HIV-RNA	
		R5	X4
PBMC	R5	78	5
HIV-DNA	X4	5	15

[a]Geno2pheno cut-off at FPR 5.75%.

PBMC of 100 HIV-infected patients [7]. It is not clear whether the HIV-infected cells in the circulation reflect the HIV tropism within the cellular reservoir. Tropism testing from PBMC samples, however, may offer further treatment options in patients with undetectable viral loads that could perhaps simplify treatment regimens and possibly lower side effects.

When genotypic assays were compared with phenotypic assays, similar sensitivities and specificities were observed, depending on the cut-offs used in the bioinformatics tools. For patient safety reasons, the geno2pheno[coreceptor] consortium implemented the "German recommendations" scheme, see Materials and methods. As new insights will emerge from future studies, the interpretation scheme for clinical use might change. Nevertheless, genotypic HIV-1 tropism assays have the advantage of short turn-around times and are less costly than phenotypic assays.

Conclusion

The sequencing assay developed for the genotypic determination of HIV-1 tropism permits analysis of a wide spectrum of HIV-1 group M subtypes, including several circulating recombinant forms. HIV-RNA from plasma or HIV-DNA from PBMC, including PBMCs from patients without detectable viral loads, can readily be assessed. The examination of 103 sample pairs of HIV-RNA from plasma and HIV-DNA from PBMC samples showed highly concordant tropism results.

With the limitation of population-based sequencing and the necessity to use a bioinformatics interpretation tool, we found the assay to be robust, accurate and suitable for the clinical routine laboratory.

References

[1] Berger EA, Dorms RW, Fenyö EM, Korber BT, Littman DR, Moore JP, et al. A new classification for HIV-1. Nature 1998;391:240.

[2] Sayana A, Khanlou H. Maraviroc: a new CCR5 antagonist. Expert Rev Anti Infect Ther 2009;7:9–19.

[3] Lengauer T, Sander O, Sierra S, Thielen A, Kaiser R. Bioinformatics prediction of HIV coreceptor usage. Nat Biotechnol 2007;25:1407–10.

[4] Harrigan PR, Geretti AM. Genotypic tropism testing: evidence-based or leap of faith? AIDS 2011;25:257–64.

[5] Kaiser R. Institute of Virology, University of Cologne, Germany; Personal communication, December 1, 2010.

[6] Verhofstede C, Vandekerckhove L, Van Eygen V, Demecheleer E, Vandenbroucke I, Winters B, et al. CXCR4-using HIV type 1 variants are more commonly found in peripheral blood mononuclear cells DNA than in plasma RNA. J Acquir Immune Defic Syndr 2009;50:126–36.

[7] Parisi SG, Androni C, Sarmati L, Boldrin C, Buonuomini AR, Andreia S, et al. A study of HIV coreceptor tropism in paired plasma, PBMC and cerebrospinal fluid isolates from antiretroviral naive subjects. J Clin Microbiol 2011;49:1441–5.

Note: The German interpretation scheme was up-dated in March 2012 and currently suggests X4 tropism at FPR $\leq$ 5%, ambiguous tropism at FPR between 5% and 15% and R5 tropism at FPR >15% (http://coreceptor.bioinf.mpi-inf.mpg.de/).

2.7 Expression of a subset of microRNAs in clinically non-functioning pituitary adenomas correlates with tumor size

Henriett Butz, István Likó, Sándor Czirják, Péter Igaz,
Márta Korbonits, Károly Rácz and Attila Patócs

Summary

MicroRNAs (miRs) are small (16–29 nt), non-coding RNA molecules that posttranscriptionally regulate protein expression via RNA interference. Pituitary adenomas are the most common intracranial tumors, but the genetic background of their pathogenesis is poorly characterized. Several previous studies, including ours, have demonstrated that altered miR expression profile determined in pituitary tumors compared to normal tissue may be involved in the pathogenic process. Our aim was to identify biological pathways that were potentially implicated in the growth of pituitary tumors, and altered by miRs. The expression profile of miRs in 8 clinically non-functioning pituitary adenomas (NFPA) and in four normal pituitary tissues was determined using miR array based on quantitative real-time polymerase chain reaction (PCR). Pathway analysis was performed by the DIANA-mirPath tool using TargetScan v.5 target prediction software followed by enrichment analysis of multiple miR target genes by Pearson's χ^2-test. Of the 457 miRs detected in both NFPA and normal pituitary tissues, 162 were significantly under- or overexpressed in NFPA compared to normal pituitary tissues. Whole array-based pathway analysis indicated alterations of several previously-suggested pathways. Expression of 18 miRs was found to be negatively correlated with tumor size. Of these 18 miRs, 6 were underexpressed in tumor tissue. Pathway analysis for these miRs revealed the involvement of Wnt, PI3K/Akt and mitogen-activated protein kinase signaling pathways. miR expression profiling is a suitable method for the identification of novel biomarkers for pituitary tumor progression. Our results support and further validate that the previously suggested signaling pathways potentially involved in pituitary tumorigenesis may be influenced by miRs.

Background

The estimated prevalence of pituitary adenomas is approximately 16%, but the clinically-relevant cases appear rarely [1]. Although they are common neoplasms, the genetic background of their pathogenesis is poorly characterized. Gene mutations are very rare, however promoter methylation and increased expression of the genes involved in regulation of the cell cycle frequently being described. Alterations of the mitogen-activated protein kinase (MAPK), PI3K/Akt, Wnt and Notch pathways were identified as being deregulated in adenomas [2].

MicroRNAs (miRs) are short, non-coding RNA molecules that posttranscription-ally regulate gene expression through the RNA interference. Their participation is well-known in development, cell proliferation and differentiation, and tumorigenesis [3]. Only a few miR–mRNA interactions have been described in the pituitary; among them downregulation of *miR-15a*, *16–1* and *let-7*, which lead to BCL2 and HMGA2 overexpression [4,5] have been reported. Our group has identified *Wee1* and *Smad3* as miR targets in pituitary adenomas [6,7].

Aim

The aim of our current work was to identify miRs and the biological pathways altered by them, which are potentially involved in pituitary tumor growth.

Methods

We worked on 12 pituitary tissue specimens, including 4 healthy control and 8 non-functioning adenoma (NFPA) specimens. Routine immunohistochemistry analysis was performed on each sample. MiR expression profile was evaluated by TaqMan MicroRNA Panel v.2 (Life Technologies, Applied Biosystems, Carlsbad, CA, USA) based on quantitative real-time PCR, including 768 miR assays. Pathway analysis and target prediction were performed using *in silico* bioinformatics tools (DIANAmiR-Path and TargetScan v.5 software, http://diana.cslab.ece.ntua.gr/pathways/).

Results

We found 162 significantly under- or overexpressed miRs in NFPA samples compared to normal pituitary. In addition, the expression level of six underexpressed miRs nega-tively correlated with tumor size.

Using target prediction and pathway analysis, we found that the three most significant pathways altered by miRs in NFPA samples were the Wnt, chronic myeloid leukemia (CML) and prostate cancer pathways which cover MAPK and PI3K/Akt signaling [7].

It was previously described that of the six investigated miRs, miR-424 and -503 both target Cdc25A, which were validated by reporter gene experiment [8]. We found that the expression of these miRs and the mRNA expression of *Cdc25a* changed in an opposite way in NFPA. Cdc25a is a phosphatase enzyme that helps G2/M transition by activating Cdk1 through its dephosphorylation. Wee1 kinase has an opposite effect, by inactivating the Cdk1-Cyclin B complex. We have previously detected that the protein level of Wee1 is decreased in both growth hormone producing and non-functioning adenomas compared to the normal pituitary. We identified and demonstrated the direct interaction between miR-128a, miR-516a-3p and miR-155 and Wee1 using bioinfor-matic and *in vitro* experimental procedures [6].

These results together may suggest that both Wee1 and Cdc25a are under miR regula-tion and their alterations may be involved in the pituitary tumorigenesis (▶Fig. 2.7.1).

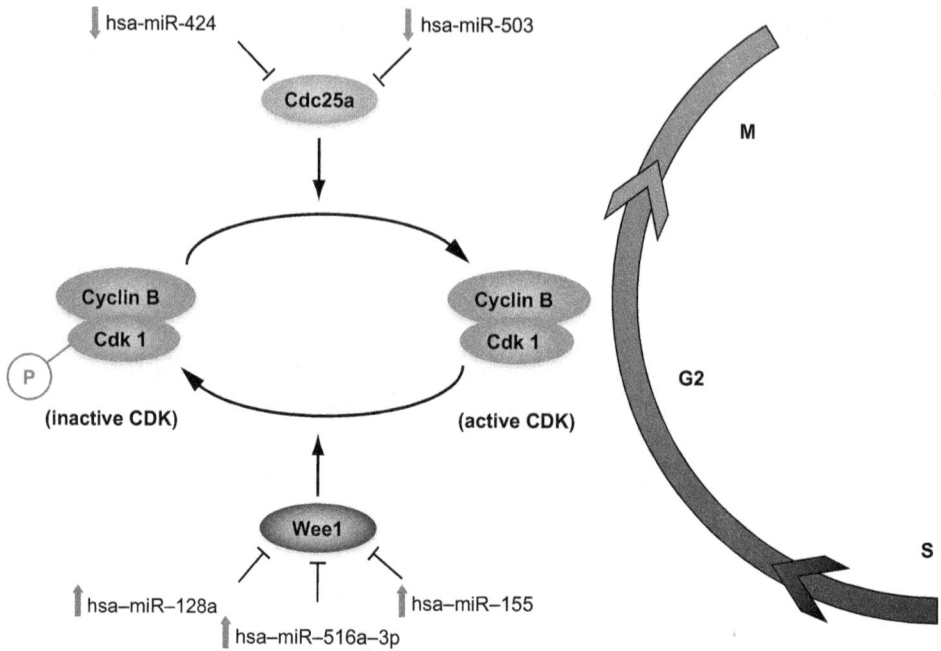

Fig. 2.7.1: Regulation of G2/M transition in non-functioning pituitary adenomas.

Conclusion

We identified six underexpressed miRs in NFPAs whose expressions negatively correlated with tumor size. The lost function of these miRs may contribute to tumor growth and progression. The decreased expression of these miRs enhances Wnt, MAPK and PI3K/Akt signaling.

MiR-424 and miR-503 target Cdc25a, and this interaction probably plays a role in the development of pituitary adenomas. Cdc25a regulates both G1/S and G2/M transition. Wee1, the antagonist of Cdc25a, is also a miR target in pituitary adenoma, hence in hormonally inactive adenomas G2/M transition is complexly regulated by Wee1 and Cdc25a, whose expression is regulated through RNA interference.

These results support and further validate the belief that the signaling pathways previously suggested to be affected in pituitary tumorigenesis may be influenced by miRs.

MiR expression profiling is a suitable method for investigating the pathogenesis of pituitary adenomas and it can be used for identifying either predictive factor of prognosis or potential therapeutic targets.

References

[1] Daly AF, Tichomirowa MA, Beckers A. The epidemiology and genetics of pituitary adenomas. Best Pract Res Clin Endocrinol Metab. 2009;23:543–54.
[2] Dworakowska D, Grossman AB. The pathophysiology of pituitary adenomas. Best Pract Res Clin Endocrinol Metab 2009;23:525–41.

[3] Krol J, Loedige I, Filipowicz W. The widespread regulation of microRNA biogenesis, function and decay. Nat Rev Genet 2010,11:597–610.

[4] Bottoni A, Piccin D, Tagliati F, Luchin A, Zatelli MC, degli Uberti EC. miR-15a and miR-16–1 down-regulation in pituitary adenomas. J Cell Physiol 2005;204:280–5.

[5] Qian ZR, Asa SL, Siomi H, Siomi MC, Yoshimoto K, Yamada S, Wang EL, Rahman MM, Inoue H, Itakura M, Kudo E, Sano T. Overexpression of HMGA2 relates to reduction of the let-7 and its relationship to clinicopathological features in pituitary adenomas. Mod Pathol 2009;22:431–41.

[6] Butz H, Likó I, Czirják S, Igaz P, Khan MM, Zivkovic V, Bálint K, Korbonits M, Rácz K, Patócs A. Down-regulation of Wee1 kinase by a specific subset of microRNA in human sporadic pituitary adenomas. J Clin Endocrinol Metab 2010;95:E181–91.

[7] Butz H, Likó I, Czirják S, Igaz P, Korbonits M, Rácz K, Patócs A. MicroRNA profile indicates downregulation of the TGFβ pathway in sporadic non-functioning pituitary adenomas. Pituitary 2011;14:112–24.

[8] Sarkar S, Dey BK, Dutta A. MiR-322/424 and −503 are induced during muscle differentiation and promote cell cycle quiescence and differentiation by down-regulation of Cdc25A. Mol Biol Cell 2010;21:2138–49.

2.8 Serological markers of gastric pathology

Alberta Caleffi

Summary

Gastric cancer remains one of the leading causes of mortality for malignant neoplasm. Due to insidious development, gastric cancer is often diagnosed at an advanced stage [1,2,3]. Serum pepsinogen I and pepsinogen II levels are known to increase in the presence of *Helicobacter pylori*-related non-atrophic chronic gastritis and the eradication of *H. pylori* is associated with a significant decrease in serum pepsinogens values. Serum pepsinogens have been used as biomarkers of gastric mucosa status, including gastric inflammation, so they might be useful for the detection of gastric neoplasm at an early phase [1,4,5,6]. We have assessed the results of serological tests for pepsinogen I, pepsinogen II, 17 gastrin and anti-*H. pylori* IgG antibodies (GastroPanel) to establish which of these markers and what association would provide more useful information on the status of gastric mucosa in patients admitted to the hospital with clinical gastroenterological symptoms.

Introduction, materials and methods

Since 2009, we have tested serological markers in patients suffering from epigastric heaviness, swelling, pain and burning. After an anamnesis interview, serum samples were collected and tested.

We evaluated the results of serological tests performed using enzyme-linked immunosorbent assay techniques (Biohit, Helsinki, Finland and Euroclone, Milan, Italy) for pepsinogen I (PGI), pepsinogen II (PGII), gastrin 17 (G17) and anti-*Helicobacter pylori* IgG antibodies (HP-IgG) (GastroPanel) from March 1, 2009 to January 12, 2011. The cut-off was established according to the recent literature [1,7,8,9], in collaboration with clinical gastroenterologists and an international database. A report was thereby prepared, containing these ranges of interpretation:

- decisional cut-off: PGI: 40–100 µg/L; and PGII: 2.5–10 µg/L;
- ratio PI/PII > 3; G17: 2.0–7.0 pmol/L; and HP-IgG: <32 IU.

Data were retrieved for 1049 tests performed on 685 women [mean age ± standard deviation (SD): 48.1 ± 8.5 years] and 364 men (mean age ± SD: 49.1 ± 18.2 years). The report shows results with the decisional range and the final interpretative comment, as defined by means of a multidisciplinary team. We mainly focused on a population with no established gastric pathology, patients with gastric mucosa inflammation or non-atrophic gastritis whether related or not with *H. pylori* infection, and patients with atrophic diffuse gastritis type or body type [3,10,11,12].

Results

The population was clustered in three main classes:

Class 1, comprising 616 patients: normal gastric mucosa (59%: mean age ± SD: 45.5±17.3 years; females/males 392/224); PGI (median, 5–95°percentiles): 67.2; 44.6–129 µg/L; PGII: 6.1; 3.1–9.5 µg/L; G17:1.4; 0.2–18.2 pmol/L; IgG–Hp: 2.7; 0.0–88.7 U/L including 361 patients with probably gastroesophageal reflux (mean age ± SD: 45.7 ± 16.4 years; females/males 218/143); PGI:64.8; 44.8–105.9 µg/L;PGII: 5.5; 2.9–9.2 µg/L; G17: 0.6; 0.1–1.8; IgG–Hp: 2.0; 0–77.4.

Class 2 (▶Tab. 2.8.1) containing 310 patients: non atrophic gastritis (PGII > 10 µg/L) (30%: mean age ± SD: 53.6 ± 16.5 years; females/males 198/112), 118 of which were *H. pylori* positive (mean age ± SD: 52.7 ± 14.7 years; females/males 76/42): PGI: 111.5; 41.5–261.3 µg/L; PGII: 16.3; 10.8–50.2 µg/L; G17: 7.8; 0.8–59.8 pmol/L; IgG-Hp: 87.2; 43.4–229.7 U/L, and 191 *H. pylori* negative (mean age ± SD: 54.1 ± 17.5 years; females/males 121/70); PGI: 145.0; 37.4–271.2 µg/L; PGII: 14.9: 10.2–53.1 µg/L; G17: 8.0; 0.2–40.0 pmol/L; IgG-Hp: 4.3; 0.03–34.6 U/L.

Class 3 (▶Tab. 2.8.2): One hundred and twenty-three patients: possible atrophic gastritis (PGI<40 µg/L) (11%: mean age ± SD: 44.4 ± 21.2 years; females/males 98/25) including 37 patients with probable atrophic gastritis of the gastric body (mean age ± SD: 56.8 ± 17.9 years; females/males 26/11); PGI: 17.7; 3.7–37.1 µg/L; PGII: 8.3: 2.7–33.5 µg/L; G17: 40.0; 7.3–84.0 pmol/L; IgG–Hp: 8.82; 0.1–100 U/L, and 86 patients with widespread atrophic gastritis (mean age ± SD: 39.1 ± 20.3 years; females/males 72/14); PGI: 33.0; 7.5–39.3 µg/L; PGII: 3.6:1.5–7.8 µg/L; G17: 0.7; 0.0–5.4 pmol/L; IgG–Hp: 2.6; 0.0–87.1 U/L.

Tab. 2.8.1: Class 2: Non-atrophic gastritis (30%) PGII > 10 µg/L: 310/1049 (HP+/HP–).

118 Pz HP + PGII > 10	
PGI	111.5 µg/L
PGII	16.3 µg/L
Ratio PI/PII	6.8
17 Gastrin	7.8 pmol/L
HP-IgG	87.2 IU

192 Pz HP – PGII > 10	
PGI	145 µg/L
PGII	14.9 µg/L
Ratio PI/PII	10
17 Gastrin	8 pmol/L
HP-IgG	4.3 IU

Tab. 2.8.2: Class 3: Atrophic gastritis (11%): 123/1049 (Diffuse/Body).

86 pz. Diffuse A.G.	
PGI	33 µg/L
PGII	3.6 µg/L
Ratio PI/PII	9.2
17 Gastrin	0.7 pmol/L
HP-IgG	2.6 IU
37 pz Body A.G.	
PGI	17.7 µg/L
PGII	8.3 µg/L
Ratio PI/PII	2
17 Gastrin	40 pmol/L
HP-IgG	8.82 IU

Discussion and conclusion

In our north Italian population, individuals clustered within each of the three subclasses need to be readdressed, with a specific diagnosis being made and treatment being started if appropriate.

The frequent occurrence of PGI levels <40 µg/L lead us to suggest that further investigations (i.e., gastroscopy) might be advisable for detecting potential atrophic changes of the gastric mucosa, which are often linked to pre-existing or chronic *H. pylori* infection or autoimmune polyendocrinopathies. An additional clinical usefulness of GastroPanel is that its results can be used to establish a more suitable plan for further invasive testing. There is growing consensus that it might be reliably used for identifying gastric involvement, and thereby for guiding the endoscopy decision-making process, in children as well [8,9].

Finally, the test might be used for assessing and monitoring populations at higher-risk for atrophic gastritis, as well as the relatives of patients affected by gastric carcinoma, rather than as a test for cancer itself according to literature [1,5]. As such, the test can contribute to improving the quality of life of patients.

References

[1] Di Mario F, Cavallaro LG. Non invasive tests in gastric diseases. Dig Liver Dis 2008; 40:523–30.
[2] Ohata H, Kitauchi S, Yoshimura N, Mugitani K, Iwane M, Nakamura H, Yoshikawa A, Yanaoka K, Arii K, Tamai H, Shimizu Y, Takeshita T, Mohara O, Ichinose M. Progression of chronic atrophic gastritis associated with Helicobacter pylori infection increases risk of gastric cancer. Int J Cancer 2004;109:138–43.
[3] Correa P, Piazuelo MB, Wilson KT. Pathology of gastric intestinal metaplasia: clinical implications. Am J Gastroenterol 2010;105:493–8.

[4] Cao Q, Ran ZH, Xiao SD. Screening of atrophic gastritis and gastric cancer by serum pepsinogen, gastrin-17 and Helicobacter pylori immunoglobulin G antibodies. J Dig Dis 2007;8: 15–22.

[5] Miki K, Urita Y. Using serum pepsinogens wisely in a clinical practice. J Dig Dis 2007;8:8–14.

[6] Ohata H, Oka M, Yanaoka K, Shimizu Y, Mukoubayashi C, Mugitani K, Iwane M, Nakamura H, Tamai H, Arii K, Nakata H, Yoshimura N, Takeshita T, Miki K, Mohara O, Ichinose M. Gastric cancer screening of a high-risk population in Japan using serum pepsinogen and barium digital radiography. Cancer Sci 2005;96:713–20.

[7] Cavallaro LG, Minelli R, Bertelè A, Franzè A, Di Mario F. Usefulness of GastroPanel for screening of atrophic gastritis in patient with autoimmune thyroid diseases. Dig Liver Dis 2008;40:528.

[8] Guariso G, Basso D, Bortoluzzi CF, Meneghel A, Schiavon S, Fogar P, Farina M, Navaglia F, Greco E, Mescoli C, Zambon CF, Plebani M. GastroPanel: evaluation of the usefulness in the diagnosis of gastroduodenal mucosa alteration in children. Clin Chim Acta 2009;402:54–60.

[9] de Angelis GL, Cavallaro LG, Maffini V, Moussa AM, Fornaroli F, Liatopoulou S, Bizzarri B, Merli R, Comparato G, Caruana P, Cavestro GM, Franzé A, Di Mario F. Useful of a serological panel test in the assessment of gastritis in symptomatic children. Dig Dis 2007;25:206–13.

[10] Wu CY, Kuo KN, Wu MS, Chen YJ, Wang CB, Lin JT. Early Helicobacter pylori eradication decreases risk of gastric cancer in patients with peptic ulcer disease. Gastroenterology 2009;137:1641–8. e1–2.

[11] Watabe H, Mitsushima T, Yamaji Y, Okamoto M, Wada R, Kokubo T, Doi H, Yoshida H, Kawabe T, Omata M. Predicting the development of gastric cancer from combining Helicobacter pylori antibodies and serum pepsinogen status: a prospective endoscopic cohort study. Gut 2005;54:764–8.

[12] Peek RM, Jr, Crabtree JE. Helicobacter infection and gastric neoplasia. J Pathol 2006;208: 233–48.

2.9 BNP as a biomarker of cardiac impairment in neonates with congenital heart diseases

Massimiliano Cantinotti, Bruno Murzi, Simona Storti and Aldo Clerico

Summary

Introduction: To evaluate the BNP assay response in neonates with different congenital heart diseases (CHDs) in the first month of life.

Materials and methods: BNP was measured in 347 neonates with different CHDs; 154 healthy children, matched for age, were used as controls. BNP was measured with a fully automated platform (Triage BNP reagents, Access Immunoassay Systems, Beckman Coulter, Inc.).

Results: BNP values were significantly higher ($p < 0.0001$) in neonates with CHD than controls (CHD patients: median 1167.5 ng/L, range 25–54,447 ng/L; controls: median 150.5 ng/L, range 5–866 ng/L). The diagnostic accuracy of BNP in differentiating neonates with CHD from controls evaluated by the area under the ROC curve was 0.88. A correlation was found between BNP values and the hours of life (ρ 0.166; $p = 0.0019$ by Spearman rank correlation analysis) that was particularly strong within the first 96 h of life (ρ 0.391; $p < 0.001$). After this initial increase in the first 4 days of life, in the following days BNP values tend to stabilize. This time course of BNP values during the first week of life differs from controls where BNP values, initially high, fall after the third day of life. Among various CHDs, defects characterized by right ventricular pressure overload showed lower BNP values ($p < 0.001$).

Conclusions: BNP showed a great accuracy in differentiating neonates with CHD from healthy subjects. BNP values showed a strong correlation with time, especially in the first 4 days of life. After this initial increase, BNP values tend to stabilize in the following days.

The measurement of B-type natriuretic peptide (BNP) and its related N-terminal pro-peptide (NT-proBNP) allowed a significant improvement in diagnostic accuracy and risk stratification in adult patients with cardiac diseases [1]. In recent years, the BNP/NT-proBNP assay has progressively gained consensus as diagnostic and prognostic biomarker in pediatric patients with congenital heart diseases (CHDs) undergoing cardiac surgery [2]. However, the lack of reliable reference values for BNP and NT-proBNP in neonates and infants has greatly limited the clinical impact of this assay and probably also produced some conflicting results. Only recently reference values calculated in large populations of healthy neonates and infants have been published for both BNP and NT-proBNP assays [2,3,4].

Diagnostic accuracy of BNP and NT-proBNP assay in pediatric patients with CHD

Recent studies [2,3,4,5] have demonstrated that measured BNP and NT-proBNP values are both age and method-dependent. BNP and NT-proBNP concentrations are very high during the first 4 days of life, with a rapid fall throughout the first week, followed by a further slower progressive reduction up to first month of life (▶Fig. 2.9.1B). After the first month of life, BNP and NT-proBNP levels show no significant changes and there are no gender-related differences up to puberty [1,2]. Furthermore, these concentrations, as well as reference ranges and cut-off values, are strictly method-dependent [2].

The diagnostic accuracy of BNP in neonates with CHD also varies during the first month of life, with the lowest diagnostic accuracy in the first three days after birth [2,3] (▶Tab. 2.9.1 and ▶Fig. 2.9.1). These findings are due to the different time courses of BNP levels observed in neonates with or without CHD (▶Fig. 2.9.1). In the first three days of life, BNP levels are very high in both healthy babies and patients, while after the fourth day peptide levels rapidly and progressively fall only in healthy neonates (▶Fig. 2.9.1). For these reasons, we suggest two different cut-off values to rule CHD in or out: the first with higher values from 1 to 3 days of life and the second with lower values from days 4 to 30 [2,3] (▶Tab. 2.9.1). Moreover, our data confirm previous findings [5,6], suggesting that BNP could be used in neonatal intensive care units to differentiate between children with respiratory distress due to cardiac defects from those with pulmonary diseases.

Clinical impact of BNP assay in patients with CHD

The introduction of a clinical routine BNP/NT-proBNP assay has significantly improved the outcome of adult patients with heart failure [7]. A similar favorable impact is theoretically expected with the introduction of the BNP/NT-proBNP assay in pediatric patients with CHD.

Even if a great variability of BNP values has been observed among diseases characterized by different hemodynamic conditions, defects presenting with right ventricular pressure overload (such as tetralogy of Fallot, pulmonary stenosis, etc.) on average showed lower BNP values than defects involving the left heart [2,3]. However, further studies are necessary to evaluate whether the BNP assay may have a role in the differential diagnosis of CHDs.

Prognostic value of BNP in children undergoing cardiac surgery for CHD

Preliminary results suggested that both basal and post-operative BNP values are accurate markers of post-operative outcome correlating with the duration of mechanical ventilation, the stay in intensive care unit and composite end-point of outcome [2]. In most of the studies performed so far, however, there was a scarce differentiation among age-subgroups and the number of neonates (if present) was extremely limited, as recently reviewed [2]. Therefore further studies, including a wider population of neonates and infants, are necessary to definitively demonstrate the prognostic role of the BNP assay in children undergoing cardiac surgery.

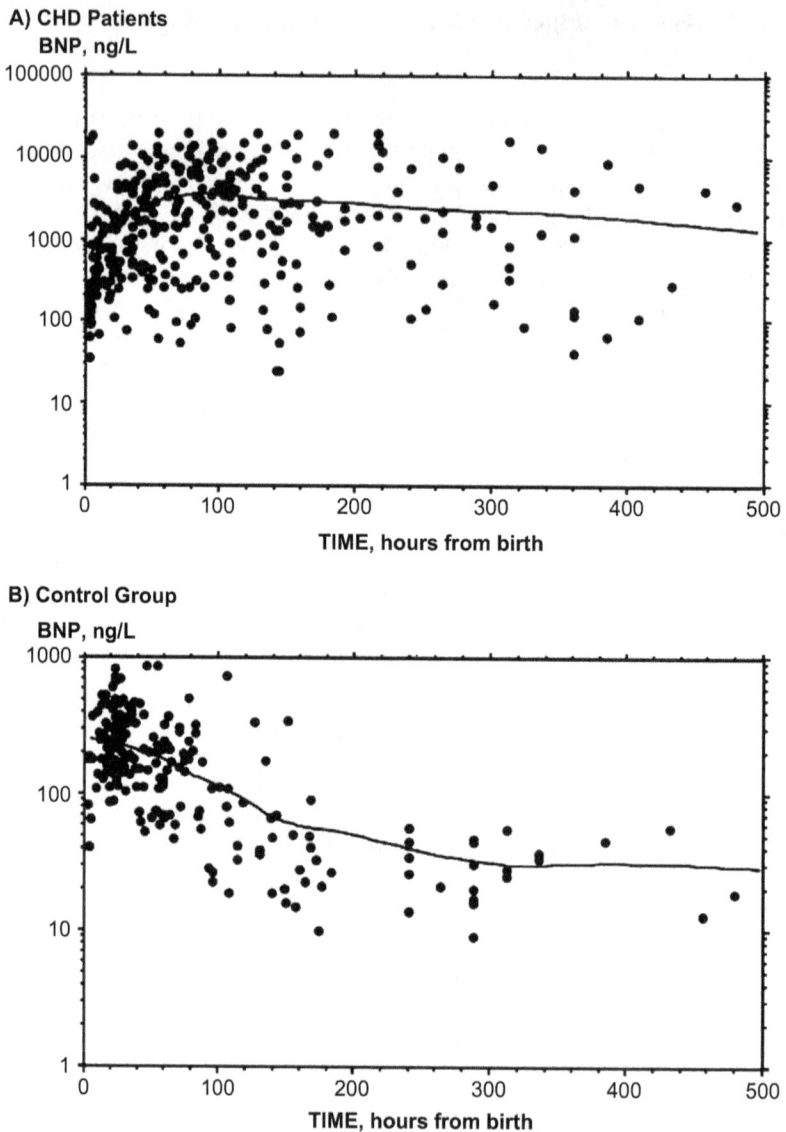

Fig. 2.9.1: Distribution of BNP circulating levels in healthy neonates and CHD patients in the first weeks of life.
(A) Distribution of plasma BNP values in neonates with CHD in the first 500 h of life.
(B) Distribution of plasma BNP values in healthy neonates in the first 500 h of life.
BNP was measured with a fully automated platform (Triage BNP reagents, Access Immunoassay Systems, REF 98200, Beckman Coulter, Inc., Fullerton, CA 92835). In the figure, the trend between BNP values and time is indicated by a continuous line. The trend was assessed by smooth (66%) spline analysis (Stat-View 5.0.1 program, 1992–98, SAS Institute Inc., SAS Campus Drive, Cary, NC, USA).

Tab. 2.9.1: AUC, best cut-off, sensitivity and specificity values of BNP assay observed in the 3 population groups, divided according to age, by means of ROC analysis.

Groups	AUC (SE)	Best BNP cut-off value, ng/L	Specificity, %	Sensitivity, %
1. Whole population (222 controls, 351 patients)	0.897 (0.013)	345	86	78
2. First 96 h of life (157 controls, 213 patients)	0.868 (0.018)	408	87	76
3. From 5 to 30 days of life (65 controls, 138 patients)	0.971 (0.011)	112	93	92

SE, standard error. Best cut-off values were calculated by ROC analysis as the BNP values minimizing the sum of false positive and negative results. Sensitivity and specificity values were calculated at the best cut-off values.

Conclusion

The introduction of age-dependent reference and cut-off values may allow a wider and correct use of the BNP/NT-proBNP assay in neonates with CHD. Studies conducted on a large population of neonates with CHDs are advised in order to understand the diagnostic and prognostic role of BNP/NT-proBNP assay in children undergoing cardiac surgery. BNP and NT-proBNP may be considered to be helpful adjunctive markers in the integrated management of neonates CHD providing independent, low cost and complementary information to conventional diagnostic imaging techniques.

References

[1] Emdin M, Vittorini S, Passino C, Clerico A. Old and new biomarkers of heart failure. Eur J Heart Fail 2009;11:331–5.
[2] Cantinotti M, Giovannini S, Murzi B, Clerico A. Diagnostic, prognostic and therapeutic relevance of B-type natriuretic hormone and related peptides in children with congenital heart diseases. Clin Chem Lab Med 2011;49:567–80.
[3] Cantinotti M, Storti S, Ripoli A, Zyw L, Crocetti M, Assanta N, et al. Diagnostic accuracy of B-type natriuretic hormone for congenital heart disease in the first month of life. Chem Lab Med 2010;48:1333–8.
[4] Nir A, Lindinger A, Rauh M, Bar-Oz B, Laer S, Schwachtgen L, et al. NT-pro-B-type natriuretic peptide in infants and children: reference values based on combined data from four studies. Pediatr Cardiol 2009;30:3–8.
[5] Ko HK, Lee JH, Choi BM, Lee JH, Yoo KH, Son CS, et al. Utility of the rapid B-type natriuretic peptide assay for detection of cardiovascular problems in newborn infants with respiratory difficulties. Neonatology 2008;94:16–21.
[6] Davlouros PA, Karatza AA, Xanthopoulou I, Dimitriou G, Georgiopoulou A, Mantagos S, et al. Diagnostic role of plasma BNP levels in neonates with signs of congenital heart disease. Int J Cardiol 2011;147:42–6.
[7] Clerico A, Fontana, M, Ripoli A, Emdin M. Clinical relevance of BNP measurement in the follow-up of patients with chronic heart failure. Adv Clin Chem 2009;48:163–79.

2.10 Are 25-hydroxyvitamin D assays fit for purpose?

Graham D. Carter

Summary

An international Quality Assessment Scheme (DEQAS) has been monitoring the performance of 25-hydroxyvitamin D assays since 1989. There has been a gradual improvement in inter-laboratory precision and most methods give mean results within about 10% of the target value (the all-laboratory trimmed mean). Observed changes in bias over time could cause problems in long-term epidemiological surveys. The measurement of 25-hydroxyvitamin D can be inconsistent, both between methods and within the same method.

Background

Methods for measuring serum 25-hydroxyvitamin D (25-OHD) have proliferated in the past few years. The more rigorous, labor-intensive extraction assays are largely being replaced with fully automated non-extraction methods that are more prone to matrix effects. Only recently has this trend been challenged, with the introduction of liquid chromatography-tandem mass spectrometry (LC-MS/MS), although this apparently highly specific method is not without its problems. In the clinical setting, the main purpose of measuring 25-OHD is to diagnose vitamin D insufficiency/deficiency or to monitor the effectiveness of vitamin D supplementation in raising the concentration of circulating 25-OHD. Ergocalciferol (vitamin D_2) is still used to treat vitamin D deficiency, so methods used in monitoring treatment should ideally be co-specific for 25-OHD$_2$. Long-term stability of assay performance is particularly important in epidemiological studies.

The vitamin D external quality assessment scheme

The vitamin D external quality assessment (proficiency testing) scheme (DEQAS) was established in 1989 (www.deqas.org). The scheme now has over 1000 participants in 40 countries. Five samples of unadulterated serum are distributed every three months and results are analyzed by the Healey method [1] to give an all-laboratory trimmed mean (ALTM) and a "robust estimator" of standard deviation (SD). The accuracy of each result is given by its percentage bias from the ALTM and the relative accuracy of each method by the percentage bias of the method mean from the ALTM.

DEQAS has investigated various aspects of 25-OHD methodology; investigations have included the effect of an anticoagulant (EDTA) and a serum-separating gel on 25-OHD results [2] and the use of a common standard in LC-MS/MS assays [3]. Blood donations used by DEQAS seldom contain 25-OHD$_2$ and the spiking of serum with 25-OHD was shown to be unsatisfactory [4,5]. Occasional samples containing endogenous 25-OHD$_2$ have been distributed.

Assay performance

There has been a reduction in overall inter-laboratory imprecision (CV) of 25-OHD results over the past 15 years, from approximately 32% in 1995 to 16% in 2010 [6]. With the exception of the Roche-25OHD$_3$ and HPLC methods (CVs 23.9% and 22.1%, respectively), the most popular assays currently have CVs below 17%. The LC-MS/MS methods, which showed considerable variability when first introduced, are now comparable to most other methods. This may be due to the wider use of a commercial standard (ChromSystems, Munich) or the National Institute of Standards and Technology standard reference materials (SRM 2972/972).

▶Fig. 2.10.1 Illustrates the change in method bias over the last 3 distribution cycles. With the exception of the IDS RIA and high-performance liquid chromatography assays, mean results for the 2010/2011 distribution were within 10% of the all-laboratory trimmed mean, although large differences exist between individual methods. Samples containing endogenous 25-OHD$_2$ have produced variable results; for a sample distributed in 2008 (25-OHD$_2$ was approx. 33% of the total), results for total 25-OHD given by immunoassay methods ranged from 76.8% (DiaSorin Liaison total) to 87% [DiaSorin radioimmunoassay (RIA)] of the LC-MS/MS method mean. A sample (with 25-OHD$_2$ being approx. 59% of the total) distributed in January 2011

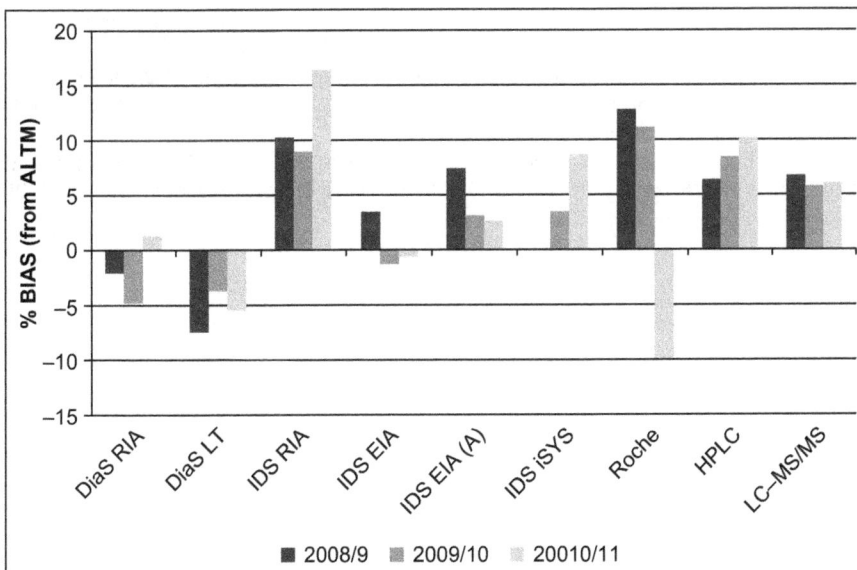

Fig. 2.10.1: Average % Bias for major 25-OHD methods in each of the last 3 DEQAS distribution cycles.
DiaS, DiaSorin; EIA, enzyme immunoassay; EIA (A), automated enzyme immunoassay; Roche, Roche 25-OHD$_3$ assay.

gave immunoassay results ranging from 63.6% (IDS RIA) to 82.1% (DiaSorin RIA) of the LC-MS/MS method mean (71.7 nmol/L). However, this sample has been analyzed by the National Institute of Standards and Technology reference measurement procedure and preliminary results indicate a total 25-OHD result of only 61.4 nmol/L – 14% lower than the DEQAS method mean. This discrepancy, so far unexplained, was not due to the presence of 3-epi-25-OHD$_3$, the concentration of which was negligible.

Conclusions

Most 25-OHD assays are probably fit for the purpose of diagnosing or confirming vitamin D deficiency, but users should be aware that the performance of assays can vary over time. A change in bias could have serious implications for long-term epidemiology studies. In monitoring supplemented subjects, an inconsistent response of assays to 25-OHD$_2$ could compromise the results of patients receiving vitamin D$_2$.

References

[1] Healey MJR. Outliers in clinical chemistry quality control schemes. Clin Chem 1979;25: 675–77.
[2] DEQAS Report, July 2009; Charing Cross Hospital, London, UK.
[3] Carter GD, Jones JC. Use of a common standard improves the performance of liquid chromatography – tandem mass spectrometry methods for serum 25-hydroxyvitamin D. Ann Clin Biochem 2009;46:79–81.
[4] Carter GD, Jones JC, Berry JL. The anomalous behaviour of exogenous 25-hydroxyvitamin D in competitive binding assays. J Steroid Biochem Mol Biol 2007;103:480–2.
[5] Horst RL. Exogenous versus endogenous recovery of 25-hydroxyvitamins D2 and D3 in human samples using high-performance liquid chromatography and the DiaSorin LIAISON Total-D Assay. J Steroid Biochem Mol Biol 2010;121:180–2.
[6] Carter GD, Berry JL, Gunter E, Jones G, Jones JC, Makin HLJ, Sufi S, Wheeler MJ. Proficient testing of 25-Hydroxyvitamin D (25-OHD) assays. J Steroid Biochem Mol Biol 2010; 121:176–9.

2.11 Update on multiple sclerosis

Anne H. Cross

Summary

Multiple sclerosis (MS) is a chronic disorder of the central nervous system (CNS) that affects 2.5 million people worldwide. Several striking new findings are notable. Although traditionally considered to be a disease primarily affecting those of Northern European ancestry, MS is increasingly being diagnosed in other populations, such as those of Middle-Eastern, African and Asian heritage. The female-to-male ratio appears to be increasing. Low serum vitamin D levels have been associated with an increased risk of developing MS. Efforts to dissect the genetics of MS using genome-wide association studies have revealed risk associations with several genes, almost all of which are related to the immune system. Damage to the CNS grey matter, as well as to white matter, is now recognized and is not uncommon. Diagnosis of MS is based on the demonstration of CNS lesion dissemination in space and time, for which no alternative diagnosis is found. Magnetic resonance images of the brain and spinal cord, often together with cerebrospinal fluid (CSF) analysis for immunoglobulin levels and the presence or absence of oligoclonal bands specific to the CSF, are critical to make the diagnosis. MS can be divided into clinical subtypes: relapsing-remitting, secondary progressive, primary progressive and progressive relapsing. In primary progressive MS, dissemination in time is revealed through increasing neurological signs over time without relapses. In the past two decades, treatments that reduce relapse rate and long-term disability have become available for relapsing clinical subtypes of MS. Treatments are expanding rapidly, with several new medications, including natalizumab, fingolimod, dimethyl fumarate, laquinimod, teriflunomide and alemtuzumab now approved or in late-phase studies.

An overview of multiple sclerosis

Multiple sclerosis (MS) is an inflammatory, demyelinating disease of the central nervous system (CNS), and is typically characterized by recurrent attacks of neurological disability. Worldwide there are an estimated 2.5 million cases of MS, of which about 70% are women. Age of onset is usually in the third or fourth decade of life. A strong relationship between geographic latitude and the risk of developing MS, both above and below the equator, has been observed, with prevalence in tropical areas only being about 1/10 of that in cooler latitudes. However, this latitudinal gradient is decreasing. MS incidence varies by race, being the highest among persons of Northern European ancestry. MS is rare among Asian populations and in the black population of Africa (but is not uncommon in African Americans).

The main genetic susceptibility locus in MS patients of Northern European ancestry is within the *HLA-DRB1* gene. Possession of the *DRB1*1501* or *DRB1*1503* haplotypes increases the risk of MS to 2–4 times that of the general population. A worldwide effort to uncover the genetic details of MS risk is ongoing [1].

A variety of infectious agents have been proposed to be related to MS development. Alotaibi et al. [2] studied pediatric MS (which is rare) and found a significantly higher rate of prior Epstein Barr virus (EBV) infection in children with MS, compared to matched control children. Eighty-three per cent of the MS group compared to 42% of the controls showed evidence of prior EBV. Additional epidemiological studies have shown a higher rate of prior EBV in adults with MS compared to adults without MS. These data do not prove a causative role for EBV.

Molecular mimicry is a candidate mechanism for linking the epidemiological and genetic findings [3] in MS pathogenesis. Lang et al. [4] demonstrated that T cells from an MS patient recognized both a *DRB1*1501*-restricted myelin peptide and a *DRB5*0101*-restricted EBV peptide.

Risk of MS is decreased with increasing serum vitamin D levels. Differences in vitamin D levels might help to explain the latitudinal differences noted in MS, because this vitamin is produced by exposure to ultraviolet light [5].

MS plaques in cortical grey matter are being increasingly recognized. These are typically smaller and less inflammatory than white matter lesions [6]. Axons are frequently lost in MS. Axon loss is believed to be the main pathological correlate of disability.

MS may be a disease of immune dysregulation. A widely-held belief is that T lymphocytes target myelin proteins, resulting in the autoimmune destruction of myelin. No specific antigen has been identified as the sole target in MS. Immune abnormalities in the cerebrospinal fluid include increased immunoglobulins with an oligoclonal pattern on electrophoresis and the presence of activated CD4+ T cells.

MS can be sub-categorized based on clinical course as relapsing-remitting (RRMS), secondary progressive (SPMS), progressive relapsing or primary progressive MS (PPMS) [7]. Most commonly, MS begins as RRMS. About half of RRMS patients subsequently develop SPMS, with progressive disability. Relapses may be superimposed on progression in SPMS. PPMS patients display progression without superimposed clinical relapses. Interestingly, the gender ratio in PPMS is 50% male and 50% female, rather than the female predominance of RRMS and SPMS.

Front-line treatments are primarily the beta-interferons and glatiramer acetate (▶Tab. 2.11.1). Among the many actions of beta-interferons is inhibition of the expression of class II major histocompatibility complex molecules, which are critical to T-cell activation. Evidence suggests that treatment with of beta-interferons delays the transition from RRMS to SPMS [8]. Glatiramer acetate is a synthetic, random polypeptide. The mechanism by which it reduces MS relapses is poorly understood, although there is evidence to suggest that it binds to major histocompatibility complex II molecules and may compete with myelin peptides for binding [9]. Other data indicate that glatiramer acetate induces a shift toward Th2 cells, thereby enhancing the populations of T cells that "regulate" the immune system [9].

Newer approved agents include natalizumab, a monoclonal antibody directed against alpha-4 integrins, which reduces relapse rate by 60%–70% in RRMS [10]. However, progressive multifocal leukoencephalopathy has occurred in over 100 patients treated

Tab. 2.11.1: Disease modifying therapies currently approved for MS.

Drug, Dosing schedule	Year Approved (USA)	Type of MS
IFN β 1b (Betaseron®) SQ every other day	1993	Relapsing forms of MS, CIS
Glatiramer acetate (Copaxone®) SQ daily	1996	RRMS, CIS[a]
IFN β 1a (Avonex®) IM weekly	1996	RRMS, CIS
IFN β 1a (Rebif®) SQ 3x/week	2002	RRMS
Mitoxantrone (Novantrone®) IV every 3 months Lifetime dose limit	2000	Worsening RRMS, SPMS, progressive relapsing MS
Natalizumab (Tysabri®) IV infusion every 4 weeks	2004/2006	Relapsing MS
Fingolimod (Gilenya®) daily oral	2010	Relapsing MS

[a]CIS, clinically isolated (demyelinating) syndrome.

with natalizumab. Fingolimod is a new oral agent that modulates expression of sphingosine-1 phosphate receptors on lymphocytes [11]. This results in retention of T and B lymphocytes in lymph nodes. As lymphocytes are not injured or deleted, immunity is retained and the cells can exit intact from the lymph nodes when the drug is stopped. There are no proven therapies to slow progression in PPMS.

New anti-inflammatory and possibly neuroprotective treatments are being studied (▶Tab. 2.11.2). Promising agents include humanized monoclonal antibodies directed against immune cell surface markers and oral immunomodulators.

Tab. 2.11.2: Therapies currently under late phase investigation for MS.

Agent	Type of MS	Mechanism of action
Cladribine	RRMS	Causes lymphocyte apoptosis
BG-12/Dimethyl fumarate	RRMS	activates Nrf2 (nuclear factor E2-related factor-2), prevents oxidative stress
Estriol	RRMS	Simulates pregnancy, neuroprotective
Laquinimod	RRMS	Th2 shift
Teriflunomide	RRMS, CIS	Inhibits dihydro-orotate dehydrogenase to block proliferation of activated T cells
Alemtuzumab	RRMS	Anti-CD52, depletes T and B lymphocytes, and monocytes
Ocrelizumab	Relapsing MS and PPMS	Anti-CD20, depletes B lymphocytes
Daclizumab	RRMS	Anti-CD25, increased CD56[bright] Natural Killer cells

Conclusion

MS is an inflammatory, demyelinating disease with relative axonal sparing. Signs and symptoms of MS vary widely among patients and may include a variety of neurological signs and symptoms. Neurological dysfunction in MS is thought to be the result of impaired conduction along partially or completely demyelinated segments of nerve fibers; however, it may also be related to inflammation and to axonal injury. Advances in brain imaging, immunology and molecular biology have increased our understanding of MS. There are several therapies currently being used for the treatment of MS, especially RRMS, but no single treatment has demonstrated dramatic efficacy in the treatment of this disease.

References

[1] International MS Genetics Consortium. Risk alleles for multiple sclerosis identified by a genomewide study. N Engl J Med 2007;357:851–62.

[2] Alotaibi S, Kennedy J, Tellier R, et al. Epstein-Barr virus in pediatric multiple sclerosis. J Am Med Assoc 2004;291:1875–9.

[3] Wucherpfennig KW. Structural basis of molecular mimicry. J Autoimmunity 2001;16:293–302.

[4] Lang HL, Jacobsen H, Ikemizu S, Andersson C, Harlos K, Madsen L, Hjorth P, Sondergaard L, Svejgaard A, Wucherpfennig K, Stuart DI, Bell JI, Jones EY, Fugger L. A functional and structural basis for TCR cross-reactivity in multiple sclerosis. Nature Immunol 2002;3: 940–943.

[5] Munger KL, Levin LI, Hollis BW, Howard NS, Ascherio A. Serum 25-hydroxyvitamin D levels and risk of multiple sclerosis. JAMA 2006;296:2832–8.

[6] Kutzelnigg A, Lucchinetti CF, Stadelmann C, Bruck W, Rauschka H, Bergmann M, et al. Cortical demyelination and diffuse white matter injury in multiple sclerosis. Brain 2005;128:2705–12.

[7] Lublin FD, Reingold SC. Defining the clinical course of multiple sclerosis: results of an international survey. National Multiple Sclerosis Society (USA) Advisory Committee on Clinical Trials of New Agents in Multiple Sclerosis. Neurology 1996;46:907–11.

[8] Trojano M, Pellegrini F, Fuiani A, Paolicelli D, Zipoli V, Zimatore GB, Di Monte E, Portaccio E, Lepore V, Livrea P, Amato MP. New natural history of interferon-beta treated relapsing multiple sclerosis. Ann Neurol 2007;61;300–6.

[9] Gran B, Tranquill LR, Chen M, Bielekova B, Zhou W, Dhib-Jalbut S, Martin R. Mechanisms of immunomodulation by glatiramer acetate. Neurology 2000;55:1704–14.

[10] Polman CH, O'Connor PW, Havrdova E, Hutchinson M, Kappos L, Miller DH, et al. AFFIRM Investigators. A randomized, placebo-controlled trial of natalizumab for relapsing multiple sclerosis. N Engl J Med 2006;354:899–910.

[11] Kappos L, Radue EW, O'Connor P, Polman C, et al. For the FREEDOMS study group. A placebo-controlled trial of oral fingolimod in relapsing multiple sclerosis. N Engl J Med 2010;362:1–15.

2.12 Microalbuminuria and urinary retinol binding protein as markers of subtle renal injury in visceral leishmaniasis: sensitivity, specificity and predictive values of the immunoturbidimetric technique

Nuha A.A. Elnojomi, Ahmed M. Musa, Brima M. Younis, Mohammed E. Elfaki, Ahmed M. El-Hassan and Eltahir A. Khalil

Summary

Leishmania donovani-related visceral leishmaniasis (VL) is endemic over large areas of Sudan. It is a serious febrile illness and is characterized by fever, hepatosplenomegaly, lymph adenopathy, pancytopenia, and renal injury. Microalbuminuria (MA) and urinary retinol binding protein (urRBP) are useful markers for glomerular and tubular dysfunctions, respectively. Paromomycin®, an amino glycoside antibiotic that is under assessment as an alternative treatment for VL, is known to be nephrotoxic. The nephrotoxicity is dose related. We report the frequency of subtle renal affection of VL and Paromomycin treatment in 46 parasitologicaly-confirmed VL patients enrolled for random treatment with different Paromomycin doses (15 mg/kg/day for 28 days or 20 mg/kg/day for 21 days) in a prospective, hospital-based and comparative study. We introduce the turbidimetric measurement for MA as a simple and field-based technique. Blood and urine were collected before and after treatment for hematological, biochemical profiles in addition to MA and urRBP measurement using competitive solid-phase, sandwich enzyme-linked immune sorbent assay (ELISA), and immunoturbidimetry. All patients (46/46; 100%) had normal serum urea and creatinine levels. More than 50% of patients had pretreatment MA detected by ELISA, whereas 54% were reactive with turbidimetry. Of the 46 patients, 4.3% had pre-urRBP detected by ELISA. Post-treatment MA was seen in more than 80% of patients who were treated with 20 mg/kg/day for 21 days Paromomycin while all of the patients who were treated with 15 mg/kg/day lost their pre-treatment reactivity. The sensitivity, specificity, positive and negative predictive values for MA using the turbidimetric technique were calculated as 100%, 86%, 85% and 100%, respectively. In conclusion, subtle renal injury in VL is mainly glomerular. Use of the 20 mg/kg/day Paromomycin should be critically investigated before implementation in routine use. Turbidimetry for MA measurement is a simple inexpensive, sensitive, and specific technique with high predictive values.

Introduction

The leishmaniases are a globally widespread group of parasitic diseases caused by a flagellate protozoa belonging to the genus *Leishmania*. The spectrum of infection with

Leishmania is widely variable and includes a diverse range of clinical manifestations including visceral leishmaniasis (VL), Post kala-azar dermal leishmaniasis, cutaneous leishmaniasis and mucosal leishmaniasis. The clinical manifestations range from the fatal form of VL to the self-limiting cutaneous leishmaniasis [1,2,3].

VL, which is caused by *Leishmania donovani*, is endemic over several parts of Sudan and is a major cause of morbidity and mortality. It is characterized by fever, weight loss, pancytopenia, hepatosplenomegaly and lymphadenopathy and can be complicated by acute renal damage [1,4]. Deposition of immune complexes and renal infiltration by infected macrophages can result in glomerulonephritis, tubulointerstitial nephritis and proximal tubulopathy [1,5,6].

Microalbuminuria (MA) is considered a useful marker of glomerulonephritis and is defined as persistent elevation of albumin of 30–300 mg/day in urine, while tubulointerstitial nephritis and proximal tubulopathy are associated with the excretion of urinary retinol binding protein (urRBP), a low molecular weight protein that is completely reabsorbed by the kidneys [7].

While sodium stibogluconate is the mainstay of treatment, recent reports of the emergence of resistance have prompted the search for alternative drugs and the use of drug combinations. Paromomycin®, an amino glycoside antibiotic that is under assessment as an alternative treatment for VL, is known to be nephrotoxic. The nephrotoxicity is dose-related [8].

Immunoassays have played vital roles in diagnosing diseases, monitoring the level of humoral immune response and identifying molecules of biological or medical interest. These assays differ in their speed, specificity and sensitivity; some are strictly qualitative, while others are quantitative. Enzyme-linked immunosorbent assay (ELISA), depends on an enzyme that is conjugated with an antibody and reacts with a colorless substrate to generate a colored reaction product. These assays are sensitive. The immunoturbidimetric technique is simple, rapid and easy to perform and is sensitive enough to detect an even slightly increased urinary albumin excretion [9,10].

This study aimed to determine possible nephrotoxic effects of VL and Paromomycin treatment and to evaluate the sensitivity, specificity and predictive values of immunoturbidimetric technique compared to standard ELISA as the gold-standard technique for the measurement of MA.

Materials and methods

Following informed consent and strict inclusion and exclusion criteria, pre- and post- treatment serum and 24-h urine samples were collected from 46 sequential and parasitologically-confirmed VL patients enrolled for random treatment with different Paromomycin doses (15 mg/kg/day for 28 days or 20 mg/kg/day for 21 days) at Kassab Rural Hospital, Gedarif State in Eastern Sudan. Urine analysis, hematological profile, serum urea and creatinine were measured for all patients. MA and urRBP were measured before and after treatment. For MA measurement, two immunological techniques were used: immunoturbidimetric and competitive solid-phase ELISA techniques. For urRBP, only Sandwich-ELISA was employed.

Results

All of the study patients (46/46; 100%) had normal serum urea and creatinine levels. The routine urine analysis showed no albuminuria or increased pus cells. However, more than 50% of patients had pre-treatment MA detected by ELISA, whereas 54% were reactive with turbidimetry. Before treatment, urRBP was detected by ELISA in 4.3% of the patients. Post-treatment MA was seen in more than 80% of patients who were treated with Paromomycin 20 mg/kg/day for 21 days, while 100% of the patients who were treated with 15 mg/kg/day lost their pre-treatment reactivity. The sensitivity, specificity, positive and negative predictive values for MA using the turbidimetric technique were calculated as 100%, 86%, 85% and 100% respectively.

Conclusion

VL in Sudan probably causes subtle renal glomerular damage in more than 50% of patients, as indicated by the presence of MA. The significance of this finding in view of normal serum urea and creatinine and good clinical response needs further verification. The presence of this renal abnormality could be attributed to the deposition of immune complexes that are mainly deposited in the glomeruli. The use of the 20 mg/kg body weight Paromomycin should be critically investigated before implementation in routine practice. The immunoturbidimetric technique is a simple, sensitive and easy-to-perform technique for the measurement of MA under field conditions.

References

[1] El-Hassan AM, Ahmed MAM, Abdul Rahim AG, Abdul Satir A, Wasfi A, Kordofani AAY, Mustafa MD, Wasfi S, Bella H, El-Hassan AM, Zijlstra EE. Leishmaniasis in Sudan. Trans R Soc Trop Med Hyg 2001;95:27–58.

[2] Musa AM, Khalil EA, Raheem MA, Zijlstra EE, Ibrahim ME, Elhassan IM, Mukhtar MM, El Hassan AM. The natural history of Sudanese post-kala-azar dermal leishmaniasis: clinical, immunological and prognostic features. Ann Trop Med Parasitol 2002;96:765–72.

[3] Zijlstra EE, Musa AM, Khalil EA, El-Hassan IM, El-Hassan AM. Post-kala-azar dermal leishmaniasis. Lancet Infect Dis 2003;3:87–98.

[4] Khalil EA, Zijlstra EE, Kager PA, El Hassan AM. Epidemiology and clinical manifestations of Leishmania donovani infection in two villages in an endemic area in eastern Sudan. Trop Med Int Health 2002;7:35–44.

[5] Van Velthuysen ML, Florquin S. Glomerulopathy associated with parasitic infections. Clin Microbiol Rev 2000;13:55–66.

[6] Efstratiadis G, Boura E, Giamalis P. Renal involvement in a patient with visceral leishmaniasis. Nephrol Dial Transplant 2006;21:235–6.

[7] Ball ST, Lapsley M, Norden AG, Cairns TD, Palmer AB, Taube DH. Urinary retinol binding protein in Indo-Asian patients with idiopathic interstitial nephritis. QJM 2003;96:363–7.

[8] Jha TK. Drug unresponsiveness and combination therapy for kala-azar. Indian J Med Res 2006;123:389–98.

[9] Tetsuo M. Use of microtiter plates for latex agglutination immunoassay of serum alpha2-macroglobulin. Ann Clin Lab Sci 2001;31:205–10.

[10] Richard AG, Thomas Jk, Barbara AO, Janis K, eds. Immunology, 5th ed. New York: W.H. Freeman and Company; 2003.

2.13 Occult hepatitis B virus infection: diagnosis and significance

Wolfram Hubert Gerlich, Christian Gisbert Schüttler and Dieter Glebe

Summary

Hepatitis B virus (HBV) infection is detected by assay of its surface antigen (HBsAg) in serum. However, many persons have HBV DNA in the liver without detectable HBsAg (occult HBV infection, OBI). OBI is often defined as HBV DNA positive, HBsAg negative in the serum, but levels of HBV DNA are too low for detection in most cases. OBI may develop after HBsAg-positive infections. OBI is more often found in patients with HIV- or hepatitis C virus (HCV)-co-infection. Partial immunity after immunization against HBV with low anti-HBs titers (<100 IU/L) favors the development of OBI, while non-immune individuals develop transient HBsAg. OBI has been connected with chronic liver disease, enhanced risk of hepatocellular carcinoma, and a worse prognosis of HCV and HIV co-infection, but this is not well established. OBI is a problem in liver transplantation and transfusion medicine. Livers of anti-HBc positive donors show severe HBV reactivation in non-immune recipients without pre-emptive antiviral therapy. Low-viremic blood donors with anti-HBc but without anti-HBs transmit frequently HBV. Thus, blood and liver donors should be screened for HBV DNA with very sensitive tests and/or for anti-HBc. The biggest problem is reactivation of HBV under immunosuppression in hematological disorders. Reactivation initially remains asymptomatic but may cause liver failure under immune reconstitution. Thus, all patients should be tested for anti-HBc and anti-HBs before immunosuppression and positive patients should be carefully monitored. A special feature of OBI is the selection of HBV mutants leading to HBeAg-negative or core promoter variants and up to 16 in 65 amino acid changes in the HBsAg loop. These mutants may be undetectable by HBsAg tests, even in high concentrations after reactivation, and they may escape the HBV-vaccine induced immunity.

Introduction

Diagnosis of hepatitis B virus (HBV) infections depends on assay of its surface antigen (HBsAg) in serum [1]. Good assays reach a detection limit <0.1 IU/mL. However, in the incubation phase, 0.1 IU HBsAg may be associated with 10^4 viruses and virtually all are infectious [2]. Due to the several week-long infectious period without HBsAg, HBV carries the largest viral risk in blood transfusion. Screening with nucleic acid amplification techniques (NATs) for HBV DNA may reduce that risk, depending on the sensitivity of the NAT, but cannot eliminate it. In rare cases, the immune system stops HBV replication before HBsAg is detectable without causing disease [3]. This happens often when vaccinated persons with low anti-HBs titers (<100 IU/L) are infected with HBV, particularly with genotypes other than A2, which is the genotype of the vaccine [4].

HBV infections without detectable HBsAg are called occult HBV infections (OBI) [5]. The great majority of OBI is persistent and develops after apparent or unapparent acute HBV when the immune system controls replication but the intrahepatic HBV DNA is not completely eliminated. Low level replication in presence of anti-HBs often leads to the selection of escape mutants with up to 16 mutations in the HBsAg loop, which can grow to high titers after reactivation.

Diagnosis of persistent OBI is possible by assay of the antibody to HBV core antigen (anti-HBc) but in 15% of OBI cases, HBV DNA is found in liver biopsies of anti-HBc negative patients [6] suggesting that the anti-HBc tests have insufficient sensitivity [7]. In some cases, anti-HBs was the only marker of OBI in non-vaccinated patients [8]. Even the most sensitive NATs are not able to detect OBI reliably because the viremia is usually too low. Detectable levels of HBV DNA (>5 IU/mL; 1 IU = ca. 5 genome equivalents, ge) are more often found in patients with HCV or HIV co-infection. It has been proposed that OBI may lead to hepatocellular carcinoma or liver disease in HIV patients, or aggravate chronic HCV, but this is not generally accepted [5].

A real problem of OBI is the potential infectivity of liver or blood donations. Livers from anti-HBc positive donors often start to replicate HBV in the immunosuppressed recipient [9]. Anti-HBc positive, anti-HBs negative blood donations may contain levels of HBV DNA detectable by individual sample NAT >20 ge/mL. Experimental infection experiments [2] and follow-up in blood recipients [10] suggest that only one in 1000 virus particles from persistent OBI are infectious. Thus, NAT-negative plasma units (>200 mL) may still transmit HBV whereas red cell concentrates (containing <20 mL plasma) do not (▶Tab. 2.13.1). The recipients usually remain asymptomatic, but if they are immunocompromised they may develop fatal liver failure.

The second problem of OBI with proven relevance is reactivation. It is rare in solid-organ transplantation or cancer therapy, but frequent in the treatment of hematological disorders that involve the depletion of B cells, e.g., by rituximab. Replication of HBV

Tab. 2.13.1: Reactivity of a NAT screening test (Ultrio plus from Novartis Vaccines & Diagnostics) with plasma donations from a donor with OBI.

Date of donation	Ultrio Plus reactive replicates tested				HBV-DNA ge/mL	RBC recipient
	Undiluted	1:4	1:6	1:8	–	
20.01.05	6/6	5/6	3/10	4/10	20	aHBc negative
23.03.05	2/6	1/6	1/10	2/10	3	aHBc/s positive
31.05.05	1/6	0/6	2/10	1/10	1	Deceased
01.08.05	4/6	3/6	6/10	5/10	8	aHBc negative
05.10.05	6/6	5/6	10/10	6/10	32	None

The donor was anti-HBc positive and anti-HBs negative but his HBsAg gene contained many escape mutations. The HBV DNA concentration was calculated by probit analysis from the number of positive replicates. The 2 red blood cell recipients receiving the highest virus load (8 and 20 ge/mL) remained anti-HBc negative suggesting absence of infectious HBV in these plasma units. The second recipient was probably positive before transfusion (data from N. Lelie and W. Gerlich, unpublished).

to high levels of viremia remains clinically silent until immune reconstitution occurs. Due to the presence of memory cells, immune pathogenesis mediated by cytotoxic T cell responses can develop very rapidly and lead to fulminant hepatitis [11]. Antiviral therapy at that stage is too late [10].

Conclusion

Occurrence of HBV in the early phase of HBV infection without detectable HBsAg is a problem in transfusion medicine that can be partially solved by sensitive NAT screening, preferably in individual donations. The number of HBsAg negative, HBV DNA positive donors will increase because vaccinated donors often develop OBI when exposed to HBV. Infectious blood or liver donors with persistent OBI can be identified by sensitive NAT or anti-HBc screening. Screening for anti-HBc is indicated in all patients before immunosuppression, although not all patients with OBI will be detected. Anti-HBc positive patients should be tested for HBsAg and if negative for HBV DNA. HBsAg or HBV DNA-positive patients should receive pre-emptive HBV therapy. The others should be monitored for HBsAg, or better for HBV DNA, and treated immediately if they turn positive. Waiting until levels of alanine aminotransferase increase is wrong.

References

[1] Schaefer S, Gerlich WH. Structure, replication and laboratory diagnosis of HBV and HDV. In: Benhamou JP, Rizzetto R, Reichen R, et al. (eds). The textbook of hepatology: from basic science to clinical practice, 3rd ed. Oxford: Blackwell Publishing; 2007, pp. 823–49.

[2] Tabuchi A, Tanaka J, Katayama K, Mizui M, Matsukura H, Yugi H, Shimada T, Miyakawa Y, Yoshizawa H. Titration of hepatitis B virus infectivity in the sera of pre-acute and late acute phases of HBV infection: transmission experiments to chimeric mice with human liver repopulated hepatocytes. J Med Virol 2008;80:2064–8.

[3] Bremer CM, Sominskaya I, Skrastina D, Pumpens P, Abd El Wahed A, Beutling U, Frank R, Fritz HJ, Hunsmann G, Gerlich WH, Glebe D. N-terminal myristoylation-dependent masking of neutralizing epitopes in the preS1 attachment site of hepatitis B virus. J Hepatol 2011;55:29–37.

[4] Stramer SL, Wend U, Candotti D, Foster GA, Hollinger FB, Dodd RY, Allain JP, Gerlich W. Nucleic acid testing to detect HBV infection in blood donors. N Engl J Med 2011;364: 236–47.

[5] Raimondo G and 27 coauthors. Statements from the Taormina expert meeting on occult hepatitis B virus infection. J Hepatol 2008;49:652–7.

[6] Raimondo G, Navarra G, Mondello S, Costantino L, Colloredo G, Cucinotta E, Di Vita G, Scisca C, Squadrito G, Pollicino T. Occult hepatitis B virus in liver tissue of individuals without hepatic disease. J Hepatol 2008;48:743–6.

[7] Huzly D, Nassal M, Vorreiter J, Falcone V, Neumann-Haefelin D, Gerlich WH, Panning M. Simple confirmatory assay for anti-HBc reactivity. J Clin Virol 2011;51:283–4.

[8] Awerkiew S, Däumer M, Reiser M, Wend UC, Pfister H, Kaise R, Willems WR, Gerlich WH. Reactivation of an occult hepatitis B virus escape mutant in an anti-HBs positive, anti-HBc negative lymphoma patient. J Clin Virol 2007;38:83–6.

[9] Blaich A, Manz M, Dumoulin A, Schüttler CG, Hirsch HH, Gerlich WH, Frei R. Reactivation of hepatitis B virus with mutated HBsAg in a liver transplant recipient receiving a graft from an anti-HBs and anti-HBc positive donor. Transfusion 2012 Feb 8. doi: 10.1111/j.1537-2995.2011.03537.x. [Epub ahead of print]

[10] Gerlich WH, Bremer C, Saniewski M, Schüttler CG, Wend UC, Willems WR, Glebe D. Occult hepatitis B virus infection: detection and significance. Dig Dis 2010;28:116–25.

[11] Westhoff TH, Jochimsen F, Schmittel A, Stöffler-Meilicke M, Schäfer J-H, Zidek W, Gerlich WH, Thiel E. Fatal hepatitis B virus reactivation by an escape mutant following rituximab therapy. Blood 2003;102:1930.

2.14 Unmet needs in chronic kidney disease testing

Graham Jones

Summary

CKD is a new acronym for chronic kidney and systematic CKD testing programs based on estimated glomerular filtration rate (eGFR) are a new phenomenon, the first national programs having started in about 2005. These programs have often been subject to continuous review to ensure ongoing quality and improvements. The aim of this paper is to consider areas where further improvements in these processes may be possible. Given the differing circumstances in different countries and regions of the world, as well as the different stages of implementation, the needs and opportunities will vary from place to place. One aspect of improvement is however the willingness to learn from each other and adopt best practice whenever possible.

CKD testing – possibilities of improving

The main components of an eGFR-based CKD testing program can be briefly summarized as follows: the right test on the right patient, which is performed, reported and interpreted correctly. In the setting of a CKD-testing program based on eGFR reporting, the key components are as shown in ▶Tab. 2.14.1. Each of the steps in the process has many possible variations and can be critically assessed for possible improvement.

The most common practice is to report an eGFR with all requests for serum creatinine [1]. This case-finding or opportunistic approach is cheap to implement and focuses on patients with a need to access healthcare. This approach has not, however, been well validated and there may be those who would benefit from testing who are missing

Tab. 2.14.1: Basic components of a National CKD eGFR testing program.

Basic components of a National CKD eGFR testing program
Report eGFR with requests for creatinine
Interpret eGFR against KDOQI CKD stages
Respond to low eGFR with:
Follow-up creatinine/eGFR
Other lab tests (depending on severity)
Clinical examination
Possible referral to kidney specialist
Education provided for doctors
Similar laboratory practices

out and those with a low pre-test probability for whom follow-up of the results of this testing is not cost-effective. The optimal processes for patient entry into the testing and management stream may benefit from further revision and inclusion of CKD testing into planned health assessment programs is in place in some countries.

Creatinine measurement remains the cornerstone of all readily available estimates of glomerular filtration rates (GFR) and thus creatinine analytical quality remains a vital issue. While major advances have been made in the alignment of routine methods to international reference standards, these are not yet global, and are not reaching their full potential. An ideal assay is accurate, precise and analytically specific. Enzymatic assays largely meet the highest quality standards but cost issues mean that Jaffe assays will be used for some time yet. Thus there remains a need to further consider assay quality standards, with the involvement of manufacturers, laboratories and quality assurance programs.

The development of formulae to estimate GFR from serum creatinine and other variables remains on-going. The work of the Chronic Kidney Disease Epidemiology Collaboration (CKD-EPI) to develop a formula using all major, suitable published studies was a landmark in international cooperation. Even so, the accuracy of this formula no better than ±30% and recent suggestions that stage 3 CKD be subdivided into 3a (GFR 30–44 mL/min/1.73 m^2) and 3b (GFR 45–59 mL/min/1.73 m^2) [2] indicates a need for improvement if patients are not to be misclassified by more than one grading.

The major unknown variable that is needed for creatinine-based GFR estimation formulae is an estimate of muscle mass. Currently formulae estimate this using age and sex, but the very limited assumption is that all patients of the same age and sex have the same muscularity. Note that this muscularity should be considered as muscle mass relative to body surface area due the nature of the modification of diet in renal disease (MDRD) and CKD-EPI formulae. There is certainly room for improvement in this aspect of GFR estimation and novel approaches may be required.

A major concern with various formulae is the possibility that different formats should be used in different racial groups. This approach is fraught with difficulties, including the assessment of race and separating the true effects of race from environmental factors, such as diet and exercise. As it is possible that the major component of racial differences is muscularity, solving the "muscle" problem indicated above may also reduce the issue of racial differences.

One issue that is of concern is the misinterpretation of eGFR in acute kidney injury where different management decisions should be taken. Improvements in the identification and awareness of acute kidney injury would have benefits for CKD testing, ensuring the two conditions are not unnecessarily confused.

Perhaps the greatest area of confusion with the introduction of routine eGFR testing has been the use of this result for drug dosing decisions. While the eGFR is "right there" on the laboratory report, it is not scaled for body size and is not yet supported by many publications in the medical literature for this purpose. The need to adjust dosages in patients with reduced GFR is of great importance and a simple approach using readily-available tools is needed. To this end, there is a need for collaboration between renal physicians, pharmacologists, laboratory technicians, drug manufacturers and government health departments to find and implement a satisfactory solution [3].

Kidney disease is predominantly a disease of adults, but this does not mean that GFR estimation in children is unimportant. The changing variables affecting interpretation of serum creatinine during growth and the lower expected creatinine concentrations make testing in children difficult and there is a need for further development. There are also other groups, such as pregnant women, where developments are needed.

Once the eGFR is reported, there is the need for it to be interpreted and appropriately responded to. The international CKD classification is under review and testing for urine albumin is likely to be part of this process [2]. The need for quality testing and reporting for albumin is also required.

Once patients with CKD have been identified, there is of course ongoing work to ensure that the best therapies are available to them.

Conclusion

CKD testing with eGFR is becoming widely adopted and is part of mainstream medicine in many countries. These developments have happened relatively recently and with remarkable speed for changes of this magnitude. All aspects of the testing process should be under constant review to ensure that this laboratory testing provides the maximum benefit for patients.

References

[1] Australasian Creatinine Consensus Working Group. Chronic kidney disease and automatic reporting of estimated glomerular filtration rate: a position statement. Med J Aust 2005;183: 138–41.
[2] Levey AS, de Jong PE, Coresh J, El Nahas M, Astor BC, Matsushita K, Gansevoort RT, Kasiske BL, Eckardt KU. The definition, classification, and prognosis of chronic kidney disease: a KDIGO Controversies Conference report. Kidney Int 2010;80:17–28.
[3] Stevens LA, Nolin TD, Richardson MM, Feldman HI, Lewis JB, Rodby R, Townsend R, Okparavero A, Zhang YL, Schmid CH, Levey AS, Chronic Kidney Disease Epidemiology Collaboration. Comparison of drug dosing recommendations based on measured GFR and kidney function estimating equations. Am J Kid Dis 2009:54:33–42.

2.15 Towards a national chronic kidney disease testing program

Graham Jones

Summary

Testing programs for chronic kidney disease (CKD) are now well established in a number of different countries. These processes were started following a National Kidney Foundation 2002 report of the Kidney Disease Quality Outcomes initiative in the United States with a recommendation that creatinine alone should not be used to assess renal function, but rather an estimate of glomerular filtration rate (GFR) based on an equation. A key stimulus for this recommendation was the development of a formula for estimation of GFR that only required parameters already commonly found in laboratory databases [1]. This report and the formula stimulated many activities that have led to major initiatives aimed at improving the detection and management of CKD. Typically these initiatives have been organized on a national level and focused on reporting of estimation of GFR. More recently other factors have been introduced, particularly the routine use of urine protein or albumin measurements as another marker of kidney damage. This article aims to provide some advice for countries or regions interested in establishing a CKD testing program, using estimate of GFR reporting as an example.

Process of establishing a national CKD testing program

The components of a chronic kidney disease (CKD) testing program are basically the same as for any pathology test. Briefly summarized, these are the right test on the right patient, and performed, reported and interpreted correctly. In the setting of a CKD testing program, the most common format is the addition of an estimate of glomerular filtration rate (eGFR) with requests for serum creatinine that are interpreted against international CKD definitions, with further investigation and management being guided by the results. An unpublished survey by the IFCC/WASPaLM Task Force on CKD has shown that national programs of this type are in place in many countries with the aim of identifying kidney disease at an earlier stage than is possible by the assessment of symptoms or effects registered on other pathology tests. Early identification allows the commencement of therapies to reduce the progression of renal disease, with a consequent reduction in the risk of death or the requirement for renal replacement therapy.

Using experience with implementing a CKD testing program in Australia and New Zealand [2] and seeking further advice from individuals involved in such programs in other countries, a number of key factors have been identified that increase the likelihood of a successful process.

The involvement of national organizations for laboratory medicine, pathology and nephrology is a very important pre-requisite. Each of these specialties has a major role to play in a successful program. It is possible that there is more than one organization in each field in a country and the involvement of all relevant organizations is strongly recommended. Specifically, the nephrology organizations are required to assess the clinical need, advise on the interpretation of results and follow-up protocols. Most importantly, they play a role in doctor education and the identification resources needed for the management of patients identified by the program. The leadership by nephrology organizations may also play a major role in ensuring wide uptake by laboratories and involvement by professional laboratory organizations. Organizations for laboratory medicine, both scientist- and pathologist-based, play a vital role in the technical requirements in terms of assay quality and implementation. For these matters, the active involvement of an external quality assurance program can be very important. So is close collaboration with the manufacturers of diagnostic equipment and reagents. Additionally, laboratories also provide a direct information conduit to all requesting doctors.

Other organizations may also play an important role, depending on local factors and the stage of the implementation. Other organizations may include government, family physicians and pharmacologists.

The process of establishing a program can be viewed in a series of stages as follows: initial planning, widespread consultation, implementation and follow-up.

The planning phase involves identifying key organizations and individuals and the establishment of broad plans for the process. A steering committee with representation from the major organizations is a useful structure. This group will guide later discussions and processes. The consultation phase involves communication and discussion of all facets of the planned program within and between all relevant groups. This is important to ensure there is agreement on key aspects and an understanding of the different viewpoints that may be held. This phase will include discussion on scientific, technical, clinical and practical issues. Implementation requires planning to ensure that all parties are prepared for their roles in the process. The most vital part of the implementation phase is education. The key target is doctors who receive the results, to ensure appropriate responses. This should take many forms, for example handouts, comments on pathology reports, an information website, training events, and so on. After implementation, education will need to be ongoing to ensure the full benefits of the process.

Follow-up after implementation is also of great importance. Ideally the implementation should be monitored to ensure widespread uptake and the effects, for example rate of referral to nephrologists, also monitored. If the process has gone well, then the follow-up will allow planning for future developments. If, however, there are unintended negative consequences, it is necessary to have a structure in place to consider any possible changes.

There are resource requirements for a CKD testing program that need to be considered. Compared to other public health measures, these costs may be very modest but success is dependent on adequate resourcing. The costs can be split into three areas: planning, implementation and patient care. The planning phase requires support for administration, meetings and associated travel. For implementation there is a need for educational processes and there will be costs for the management of additional patients who are identified to have CKD.

There are a range of aspects that may need to be considered in developing a national program, especially if there are differences in practice in the region. An example list for an eGFR program is shown in ▶Tab. 2.15.1. A key issue is that the decisions are taken together, in keeping with the local needs and ownership of the processes. Inputs to decision-making can include scientific journals, results from local quality assurance

Tab. 2.15.1: Items for consideration in developing a National eGFR testing Program for CKD.

Items for consideration in developing a National eGFR testing Program for CKD
Creatinine assay quality – traceability to IDMS and precision
Units for reporting serum creatinine and GFR
If changes to creatinine values are planned, also implement changes to creatinine reference intervals
Limitations to be applied to reporting eGFR: age, dialysis, pregnancy, inpatients, only upon request
Choice of formula for calculating eGFR
Agreed responses to eGFR values and supporting educational material
Role of eGFR for drug dosing decisions
Decision on influence of race on the results
Methods of education delivery

programs and suggestions from key opinion leaders. There is also advice and support to be obtained from other groups, with the National Kidney Disease Education Program website in the US providing an excellent resource.

It is important to identify possible inhibiting factors. These may include limitations in the pathology information technology systems, lack of acceptance from important groups, inadequate resources for implementation and lack of suitable quality assays. It is also important to recognize the follow-on effects of any actions. For example, if creatinine assays require re-standardization then reference intervals and other clinical guidelines may require adjustment.

I support the advice from a developer of one national program describes the process best: "communicate, communicate, communicate, then educate".

Conclusion

CKD testing programs as described above are an example of the appropriate management of pathology testing [3]. The collaborative efforts of clinicians, pathologists and laboratory scientists are harnessed to improve patient health. It is hard to envisage a successful program without this collaborative approach. Successful collaborations in this field can also provide a model for similar processes in other fields of laboratory testing.

References

[1] Levey AS, Bosch JP, Lewis JB, Greene T, Rogers N, Roth D. A more accurate method to estimate glomerular filtration rate from serum creatinine: a new prediction equation. Ann Intern Med 1999;130:461–70.
[2] Mathew TH, Johnson DW, Jones GR. Chronic kidney disease and automatic reporting of estimated glomerular filtration rate: revised recommendations. Med J Aust 2007;187:459–63.
[3] Mathew T. Chronic Kidney Disease – An exemplar for collaboration between the clinic and the laboratory. Clin Biochem Rev 2011;32:51–3.
[4] National Kidney Foundation. K/DOQI Clinical Practice Guidelines for Chronic Kidney Disease: part 5. Evaluation of Laboratory Measurements for Clinical Assessment of Kidney Disease. Guideline 4. Estimation of GFR. Am J Kidney Dis 2002;39(S1):S76–S110.

2.16 Biochemistry and metabolism of vitamin D

Glenville Jones

Summary

Vitamin D_3 still commands a lot of scientific attention 40 years after its metabolism and its importance in calcium and phosphate homeostasis were revealed. Part of this new-found interest stems from the discovery that its active form, $1\alpha,25$-dihydroxyvitamin D_3 ($1\alpha,25$-$(OH)_2D_3$), through its nuclear vitamin D receptor, regulates hundreds of genes around the body including those involved in cell differentiation and cell proliferation. Furthermore, epidemiological association studies have suggested that levels of the main circulating form, 25-hydroxyvitamin D_3 correlate positively with various health outcomes connected to major diseases: cancer, immune function and infections and cardiovascular disease. Consequently, the biochemistry around the metabolism of vitamin D, its mechanism of action in target cells and the clinical chemistry around its specific and sensitive assay remain relevant. This short review will discuss the current state of knowledge of the cytochrome P450-enzymes involved in activation and inactivation of vitamin D, as well as provide a synopsis of the biochemistry and physiology surrounding its roles in the body. The article will end with some of the pertinent questions about vitamin D for the clinical chemist: what, how, when and in whom to measure?

Biochemistry and metabolism of vitamin D

Vitamin D_3 is formed by ultraviolet B (290–315 nm) irradiation of the skin 7-dehydrocholesterol (►Fig. 2.16.1). Transport of vitamin D_3 from skin to the storage tissues or liver for the first step of activation is carried out by a specific vitamin D binding globulin. D vitamins can also be derived from the diet, as vitamin D_3 and vitamin D_2. Vitamin D_2 and D_3 are activated and function in the same way [1], so this review will focus on vitamin D_3.

Vitamin D_3 is first activated by 25-hydroxylation in a step that is probably catalyzed by the liver cytochrome P450, CYP2R1. CYP2R1 is the best candidate for 25-hydroxylase involvement since only CYP2R1 is able of hydroxylate vitamin D_2 or vitamin D_3 equally; a human mutation at Leu[99]Pro in CYP2R1 results in rickets and the enzyme has been successfully crystallized with its vitamin D substrate bound in the active site [2]. A recent genome-wide association study [3] implicated CYP2R1 as one of the 4 major genetic determinants of serum 25-hydroxyvitamin D (25-OH-D_3), the others being: vitamin D binding globulin, CYP24A1 and 7-dehydrocholesterol reductase.

25-OH-D_3 is converted by 1α-hydroxylation to the active form of vitamin D, $1\alpha,25$-$(OH)_2D_3$ [4], primarily in the kidney, by the action of the mitochondrial cytochrome P450, CYP27B1. Synthesis of circulating $1\alpha,25$-$(OH)_2D_3$ in the normal, non-pregnant mammal occurs exclusively in the kidney, since patients with chronic kidney disease exhibit low renal-1α-hydroxylase enzyme activity and greatly reduced serum 1α,25-(OH)$_2$D$_3$ levels. Renal CYP27B1 is up-regulated by parathyroid hormone (PTH) as part of the calcium homeostatic loop and down-regulated by fibroblast-like growth factor 23 (FGF-23) as part of phosphate homeostatic loop (see ►Fig. 2.16.1). CYP27B1 protein can be detected in extra-renal tissues and this has led to the idea that extra-renal CYP27B1 exists to promote "local" production of 1α,25-(OH)$_2$D$_3$ from 25-OH-D$_3$ in tissues such as skin, prostate, intestine, breast and possibly bone [5].

The inactivation of 25-OH-D$_3$ and 1α,25-(OH)$_2$D$_3$ involves another cytochrome P450 known as CYP24A1, and formerly known as 24-hydroxylase. CYP24A1 is a multi-catalytic enzyme responsible for a 5-step, vitamin D-inducible, C-24-oxidation pathway which inactivates the vitamin D molecule, changing it to a water-soluble biliary form, calcitroic acid. Recently [6], mutations of CYP24A1 have been implicated as a cause of idiopathic infantile hypercalcemia, the hypercalcemic symptoms of which resemble those seen in the CYP24A1-knockout mouse.

The hormonal form, 1α,25-(OH)$_2$D$_3$, whether formed in the kidney or locally, has calcemic roles that include the regulation of blood calcium and phosphate concentrations by actions at the intestine, bone, parathyroid and kidney. It also has non-calcemic roles that include cell differentiation and anti-proliferative actions in various cell types: bone marrow (osteoclast precursors and lymphocytes), immune system, skin, breast and prostate epithelial cells, muscles and intestine [7]. 1α,25-(OH)$_2$D$_3$ does this through a vitamin D receptor (VDR)-mediated transcriptional mechanism, in which the hormone directly regulates gene expression of a wide variety of vitamin D-dependent genes in vitamin D-target cells. Liganded VDR recruits a partner called the retinoid X receptor and a plethora of other transactivators, termed a vitamin D receptor interacting protein (DRIP) complex, in order to transactivate genes [8].

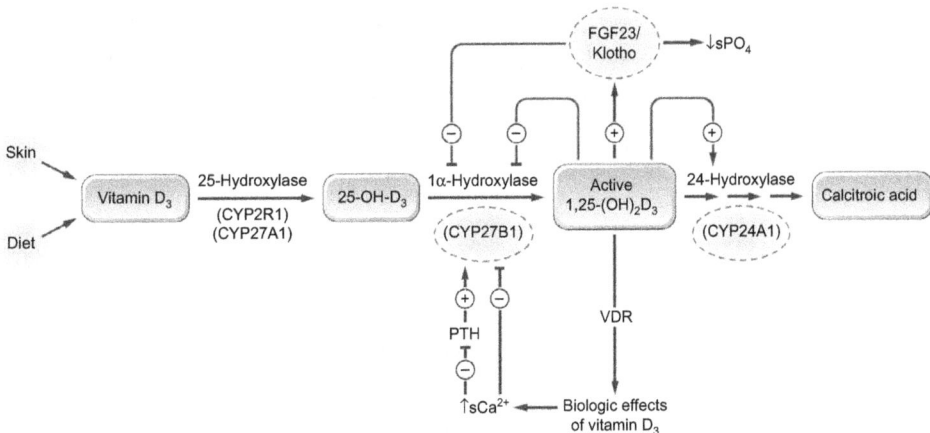

Fig. 2.16.1: Vitamin D metabolism and some physiological actions.
Current knowledge of the cytochrome 450-containing enzymes in the activation and inactivation of vitamin D together with some information about the physiological actions and regulation of levels of 1α,25-(OH)$_2$D$_3$. Reproduced with permission from NEJM [6].

Implications for the clinical chemist

What should we measure? Serum 25-OH-D$_3$, which is the sum of 25-OH-D$_2$ plus 25-OH-D$_3$, represents the best measure of the vitamin D status of the animal *in vivo* and this parameter correlates well with several clinical outcomes. Serum 1α,25-(OH)$_2$D is only useful clinically in sarcoidosis, idiopathic infantile hypercalcemia and metabolic bone diseases.

How should we measure 25-OH-D? There are now a wide range of antibody-based (kit) and liquid chromatography tandem-mass spectrometry (LC-MS/MS) based (service) methods available and their performance is discussed in the paper by Graham Carter [9].

When should we measure 25-OH-D? This parameter undergoes a seasonal fluctuation with a minimum in February/March and a maximum in September/October in northern latitudes. Vitamin D supplementation can usually be followed by monitoring on a quarterly basis, since 25-OH-D has a half-life of 20 days and serum levels take 3–4 months to plateau.

Who is at risk of vitamin D deficiency? The Institute of Medicine [10] recently defined vitamin D deficiency as serum 25-OH-D < 50 nmol/L and identified several high-risk groups including individuals who avoid sun exposure, those at high latitude, those with darkly-pigmented skin, those who are obese and those who suffer from chronic kidney disease. In these times of rising healthcare costs, monitoring of serum 25-OH-D should be focused on these high-risk groups.

Conclusions

Vitamin D remains a hot topic because its functions around the body are so general and its disease implications are still hotly debated. There is still a large concern that segments of our population are at high-risk of vitamin D deficiency and thus monitoring of serum 25-OH-D remains a central focus for clinical chemists. These assays are improving and their use is being directed at those suspected of being vitamin D deficient rather than the assay being used as a routine screening tool. Time will tell whether this is the correct decision.

References

[1] Jones G, Strugnell S, DeLuca HF. Current understanding of the molecular actions of vitamin D. Physiol Rev 1998;78:1193–231.

[2] Strushkevich N, Usanov SA, Plotnikov AN, Jones G, Park H-W. Structural analysis of CYP2R1 in complex with vitamin D3. J Mol Biol 2008;380:95–106.

[3] Wang TJ, Zhang F, Richards JB, Kestenbaum B, van Meurs JB, Berry D, Kiel DP, Streeten EA, Ohlsson C, Koller DL, Peltonen L, Cooper JD, O'Reilly PF, et al. Common genetic determinants of vitamin D insufficiency: a genome-wide association study. Lancet 2010;376:180–8.

[4] Jones G, Prosser DE. The activating enzymes of vitamin D metabolism (25- and 1α-hydroxylases). In: Feldman D, Pike W, Adams J, eds. Vitamin D, 3rd ed. San Diego: Elsevier; 2011, pp. 23–42.

[5] Holick MF. Vitamin D deficiency. N Engl J Med 2007;357:266–81.

[6] Schlingmann KP, Kaufmann M, Weber S, Irwin A, Goos C, John U, Misselwitz J, Klaus G, Kuwertz-Bröking E, Fehrenbach H, Wingen AM, Güran T, Hoenderop JG, Bindels RJ, Prosser DE, Jones G, Konrad M. Mutations of CYP24A1 and idiopathic infantile hypercalcemia. N Engl J Med 2011;365:410–21.

[7] Haussler MR, Haussler CA, Whitfield GK, Hsieh JC, Thompson PD, Barthel TK, Bartik L, Egan JB, Wu Y, Kubicek JL, Lowmiller CL, Moffet EW, Forster RE, Jurutka PW. The nuclear vitamin D receptor controls the expression of genes encoding factors which feed the "Fountain of Youth" to mediate healthful aging. J Steroid Biochem Mol Biol 2010;121:88–97.

[8] Dowd DR, MacDonald PN. Coregulators of VDR-mediated gene expression. In: Feldman D, Pike W, Adams J, eds. Vitamin D, 3rd ed. San Diego: Elsevier; 2011, pp. 193–209.

[9] Carter G. 25-hydroxyvitamin D assays: are they fit for the purpose. Proceedings of IFCC-WorldLab Meeting Berlin, May 2011. Berlin: DeGruyter, in press.

[10] Ross AC, Manson JE, Abrams SA, Aloia JF, Brannon PM, Clinton SK, Durazo-Arvizu RA, Gallagher JC, Gallo RL, Jones G, Kovacs CS, Mayne ST, Rosen CJ, Shapses SA. The 2011 report on dietary reference intakes for calcium and vitamin D from the Institute of Medicine: what clinicians need to know. J Clin Endocrinol Metab 2011;96:53–8.

2.17 Diagnostics of thalassemia

Martha Kaeslin, Saskia Brunner-Agten and Andreas R. Huber

Summary

Hemoglobinopathies are a group of heterogenic genetic diseases in which the production of hemoglobin is negatively affected. Hemoglobinopathies show up either as defects in hemoglobin quality resulting in abnormal hemoglobin structure or in a quantitative reduction of hemoglobin synthesis, known as thalassemias. Hemoglobinopathy is one of the most common monogenetic diseases in the world. Three per cent of the global community are carriers. In Switzerland about 100,000 people are affected, of which 200–300 show severe forms. In this article we will focus on thalassemias and thalassemic hemoglobinopathies. The most common forms are α- and β-thalassemias, depending on the defect in the α- or β-hemoglobin gene. α-Thalassemias are usually caused by deletions; whereas β-thalassemias are commonly induced by mutations. Based on the deletion size of the α-globin chain gene, α-thalassemias are classified as α^+ or α^0. If just one gene is inactivated, we talk about α^+ thalassemia; if the deletion size is larger, α^0 thalassemia is the result. The original distribution of hemoglobinopathologies was manifested as a belt over Africa, the Mediterranean region, the Middle East and the Indian subcontinent throughout Southeast Asia, based on the malaria distribution and the selective advantage of hemoglobinopathies within malaria-endemic regions. Screening and diagnosis of hemoglobinopathies can be carried out in specialized labs, but the analytical methods used are quite different. These need to be discussed and standardized because new molecular biological methods give the possibility of getting reliable and cost-effective results. From this point of view, it also should be kept in mind that diagnostic expenses are cheaper than treatments resulting from incorrect diagnosis. A definitive diagnosis allows proper care of the patients and genetic counseling of affected individuals.

Introduction

Hemoglobinopathy comprises entities that are generated by either abnormal hemoglobin or thalassemias. While abnormal hemoglobin is caused by a qualitative structural abnormality of the hemoglobin molecule, thalassemias result in a diminished synthesis of the globin chain [1]. Common abnormal hemoglobin like in Sickle cell anemia (hemoglobin S or HbS) is mostly found in central Africa, correlating to regions where malaria is endemic. In Africa around 7% of the population carries HbS; whereas in Southeast Asia around 4% present with hemoglobin E (HbE); and in Western Africa hemoglobin C (HbC). Overall, HbS, HbE and HbC are prevalent in about 4% of the population [2].

Due to increased immigration from Asia, Africa and the Mediterranean to Northern Europe, hemoglobinopathies like HbS, HbC and HbE are also encountered quite commonly in Switzerland. Although other abnormal hemoglobins are rare, they can cause clinically relevant symptoms. This includes hemolysis, polyglobulia, cyanosis or a combination thereof. In this article we will focus on thalassemias that are divided into α- and β-thalassemias, depending on which hemoglobin gene is affected. α -Thalassemias are divided into α^+ thalassemias, where only one α-gene is deleted. The most prevalent deletion, accounting approximately 80%, is the $-\alpha^{3.7}$ deletional form. A α^0 thalassemia is present if a larger deletion on the α-chain is present and more than one gene is affected, resulting in complete absence of hemoglobin synthesis. The symptoms of α-thalassemias can be quite diverse, from an asymptomatic carrier state to a hydrops fetalis. The classification of α-thalassemias is shown in ▶Tab. 2.17.1 and mainly depends on how and how many alleles of the α-chain gene are affected.

Thalassemia syndromes with 2 million affected individuals are the most prevalent monogenetic diseases worldwide. Due to immigration to Switzerland, thalassemia syndromes are also found quite commonly among our patients (10%–15% of all hypochromic, microcytic, anemias), coming second only to iron deficiency. Importantly, thalassemias and hemoglobinopathies can occur concomitantly, sometimes even with a normal hemoglobin variant. This results in highly variable presentations, making diagnosis from clinical observation and routine laboratory tests difficult. Therefore a stepwise diagnostic algorithm including the selective use of molecular biological methods is recommended [3].

Depending on whether the molecular defect is affecting the expression of the α- or β-chain, differentiation between α- or β- thalassemia [4] is made. α-Thalassemias are commonly based on deletions, whereas β-thalassemias are more often caused by mutations. α-Thalassemias are divided into α^+ or α^0 thalassemias, depending on the deletion size in the α-region on chromosome 16. Within a β^0 thalassemia there is a complete failure of gene activity. A β^+ thalassemia is present if there is some residual hemoglobin synthesis.

Methods

The molecular hemoglobinopathy diagnosis should follow specific diagnostic algorithms employing different laboratory methods. These methods include:

- hybridization assays;
- polymerase chain reactions (PCRs)
- multiplex-amplification refractory mutation system;
- multiplex ligation-dependent probe amplification; and
- oligonucleotide microarray chip diagnostics.

Tab. 2.17.1: Genotype and phenotype in α-thalassemia diagnostics.

Genotype	Diagnosis	Clinic
αα/αα	Heterozygous	Normal
-α/αα	Heterozygous α^+ thalassemia	Thalassemia minima
-α/-α	Homozygous α^+ thalassemia	Thalassemia minor
-/αα	Heterozygous α^0 thalassemia	Thalassemia minor
-/-α	Compound heterozygous	Hemoglobin H disease
-/$\alpha^T\alpha$	α^0/α^+ thalassemia	Hemoglobin H disease (severe)
-/-	Homozygous α^0 thalassemia	Hydrops fetalis

Tab. 2.17.2: Advantages and disadvantages of different α-thalassemia diagnostic methods.

Method	Interpretation	Handling	Costs	Occurrence
Hybrid	Limited	Relatively simple	Cheap	Common
M-PCR	Limited	Laborious	Costly	Common
RT-PCR	Extensive	Relatively simple	Efficient	New
MLPA	Limited	Skills	Expensive	Common
Arrays	Extensive	Skills	Expensive	New
Deep sequencing	Fully extend	Laborious	Very expensive	Very new

PCR, polymerase chain reaction; M-PCR, multiplex PCR; RT-PCR, real-time PCR; MLPA, multiplex ligation-dependent probe amplification.

Additionally, a real-time PCR for detailed diagnosis of α-thalassemia and thalassemia-screening has recently been patented and may be efficiently used in the near future in routine laboratory settings (data not published). The advantages and disadvantages of the various methods are shown in ▶Tab. 2.17.2.

Conclusion

Defects in hemoglobin genes are quite common and their appearance is very hetero-geneous. However, nowadays molecular techniques allow a definitive diagnosis and genetic counseling. A rational approach is taken, depending on the prevalence and the ethnic background of a patient. Targets for a molecular biological diagnosis include α-thalassemias with large deletions, severe patient symptoms, β-thalassemias with confusing clinical presentation and discrepant test results, combined conditions of thalassemia and hemoglobin variants or a post transfusion situation.

Regarding costs it has to be kept in mind that diagnostic expenditures are cheaper than treatments. It also should be mentioned that there is a difference between screening and a confirmational diagnosis. Screening assays should be cheap, sensitive and robust. The results of screening assays – if positive – need to be confirmed by very specific tests in specialized laboratories using different kinds of assays. In the future deep sequencing methods and array technologies may prove promising once they become affordable.

Last but not least next to the laboratory work, counseling and patient care are the most important factors in patient handling and should not be neglected.

References

[1] Kleinhauer E, Kohne E, Kulozik AE. Anormale Hemoglobine und Thalassemie Syndrome: Grundlage und Klinik. Landsberg: Ecomed; 1996.
[2] Huber AR, Ottiger C, Risch L, Regnass S, Hergesberger M, Herklotz R. Hämoglobinopathien: Pathophysiologie und Klassifizierung. Schweizerisches Medizinisches Forum 2004;4:895–901.
[3] Herklotz R, Risch L, Huber AR. Hämoglobinopathien-Klinik und Diagnostik von Thalassämien und anormalen Hämoglobinen. Therapeutische Umschau 2006;63:35–46.
[4] Weatherall DJ, Clegg JB. The thalassemia syndromes. 4th ed. Oxford: Blackwell Scientific Publications; 2001.

2.18 The specific roles of assessors during accreditation

Wim Huisman

Summary

Accreditation of competence means that a third-party can be sure that the laboratory fulfils all the requirements stated in a standard. For medical laboratories, this is the ISO 15189:2007. In Europe the only institutes that are allowed to perform accreditation are the national accreditation bodies who are members of the European co-operation for accreditation. Within this organization, the Health Care Committee tries to harmonize the way in which assessment is carried out to ensure that mutual recognition in practice is accomplished. The primary function of the lead assessor is to judge the quality system. The different items are presented in Chapter 3 of ISO 15189. Points for attention are: continuous quality improvement, the Plan-Do-Check-Act (PDCA) cycle, working internal audit, documentation, complaints system, management review, etc. Knowledge about quality systems is essential, although familiarity with medical laboratories is also needed. The technical assessor needs to be a medical laboratory specialist and competent in the scope that is assessed as well. Their role is primarily to judge the competence of the laboratory as given in Chapter 4 of ISO 15189. Not only related to the examination aspects as validation of methods, traceability, internal and external quality control, up to date methods and adequate standard operating procedures, turn-around time, but also the pre-examination aspects, even if the phlebotomy is not done by the laboratory, and post-examination aspects, especially the consultancy function of the laboratory specialists. Part of the assessment is the competence of laboratory technicians and the laboratory specialist. This is in relation to their number, specialties in extensive laboratories, their continuing education and renewal of registration. Contact with the physicians who order the tests should also be evaluated. A combined responsibility if the assessment team is to establish confidence in the reliability of the service offered by the laboratory. This means not just the presence of competent people, but real cooperation between them. Many inconsistencies are not just mistakes, but are related to deficiencies in the system. Re-registration of laboratory professionals is not the competence of accreditation bodies. Registration of attendance in congresses, post doctoral education and publications is helpful, but assessments can be used for this as well. The route for the decision and the responsibility is different. However, there is an overlap between both aspects, and the information from one is helpful for the other. Structured training of laboratory professionals to become assessors is needed for both aspects.

Accreditation process

Accreditation is a process whereby an organization performing medical laboratory testing can demonstrate its competence by conforming to the ISO 15189 [1] and being able to show the national accreditation body (NAB) it does. The NAB conforms to ISO 17011: 2004 [2]. Well-trained medical laboratory professionals are essential to meet the competence standards.

According to the principles of the International Federation of Clinical Chemistry, which are also mentioned in the EA-4/17;2008 "EA position paper on the description of scopes of medical laboratories," the majority of the overall service provided by the laboratory should be covered by the scope. It is recognized that the NABs cannot enforce it, but they will encourage medical laboratories to cover the majority if their examinations within each medical field in their scope. The first level of flexibility can be defined as a medical field or discipline, such as for example clinical chemistry, hematology, immunology, etc. At national level, there may be differences in the way the NAB and corresponding medical professionals define this discipline.

For each medical field mentioned in the scope, it is expected that the laboratory provides a full service, which includes all pre-examination, examination and post-examination aspects that are essential for providing an effective and efficient laboratory service for patients. Within this, it is expected that a medical laboratory is able to demonstrate its competence in interpreting the results of the examinations performed. This is in line with the intention of Working Groups (WG) on accreditation and ISO/CEN standards of European Federation Clinical Chemistry and Laboratory Medicine (EFCC) as published [3].

At European level, the NABs cooperate within the European co-operation for Accreditation (EA). It is important to realize that by European law the NABs are the only bodies that can accredit according to ISO standards. These NABs can arrange for certificates of accreditations to also be accepted elsewhere (through a Mutual Lateral Agreement), which corresponds to the Mutual Recognition Agreement of International Laboratory Accreditation Cooperation.

In the EA, the Health Care Committee, with members from European Diagnostic Manufactured Association, NABs (from more than 20 countries) and professionals (WG Accreditation from the EFCC), tries to harmonize the way the assessment is performed.

The competence of the AB is essentially based on its staff, the assessors, experts and committees. The assessment is the core activity of the AB.

In the opinion of the members of the WG accreditation, confidence in the competence of assessors is essential. A laboratory can only profit from accreditation if the assessment is done properly. A great deal of attention to minor non-conformities can be really counterproductive.

Guidelines on the qualification and competence of assessors were provided by the International Laboratory Accreditation Cooperation and EA years ago [4]. This was updated in ILAC – G11:07 [5]. What matters is how it functions in practice.

Looking to ISO 15189, the management requirements, as mentioned in Chapter 3, are the primary responsibility of the lead assessor. Important points are: continuous quality improvement in the PDCA cycle, function of the internal audit system, documentation, function of the complaint system and a functioning management review cycle. Knowledge about these quality aspects in a medical laboratory is essential. The performance of more than 5 assessments a year is very helpful and good steering of the assessment team is essential.

For the technical assessor we need a "peer" in the field: a respected well trained laboratory professional with real practical experience in the specific laboratory field that is being assessed. The focus is on the points mentioned in Chapter 4 of ISO 15189.

It is about the state of the art of the tests, the way they are performed (Internal Quality Control, External Quality Control), their validation (with attention on interference and traceability) and also on the consultation function from the laboratory professional. Sufficient contacts exist in the hospital (if the laboratory is a hospital lab), with primary care physicians (if these are their primary clients) or with other laboratory colleagues (for a referral laboratory).

In the training of the technical assessor, another point of attention needs to be their knowledge of the quality system. Are they aware of the PDCA cycle, have they got experience in seeing a broader perspective? Are they aware that a small mistake does not signify that the system is not in order, or vice versa? And do they have the necessary "soft skills" as: open-mindedness, diplomacy, being observant, perceptiveness, persistence, integrity, ability to work in a team. To have good cooperation as a team is important, and in this part the lead assessor has to play the most important role.

Even if the laboratory does not do the phlebotomy itself, the pre-analytical aspects should receive attention. A perfect analysis of a sample that is compromised in the pre-analytical phase (either in taking the sample, handling or transportation) does not contribute to the good handling of patients.

In relation to the consultation function of the laboratory specialist, attention will be paid to registration, but just registration is not enough. Points can be gathered by just being present, or in fields that are not really related to the specific aspects of the laboratory.

It is an important task of professional societies to ensure a system to monitor the adequacy of this function in all the laboratories. Well trained assessors can play a role, but this in relation with their professional societies, more than the NABs.

This means there is a combined responsibility of the NAB and the professional society in ensuring the quality of the medical laboratory service.

Conclusion

Harmonization of assessment between countries is needed. Special attention needs to be given to the way in which the technical assessors function in practice. They have an important but difficult role, especially in relation to the consultation function of medical laboratories. Quality in this field is a combined responsibility of the NABs and the laboratory societies. Training of assessors can be helpful. Learning from each other on a European scale is possible by good cooperation between the EA Health Care Committee of and the EFCC's WG on Accreditation.

References

[1] ISO 15189: Medical laboratories – particular requirements for quality and competence, ISO, 2007.
[2] ISO/IEC 17011: Conformity assessment – general requirements for accreditation bodies accrediting conformity assessment bodies, ISO, 2004.
[3] Huisman W, Horvath AR, Burnett D, Blaton V, Czikkely R, Jansen RT, et al. Accreditation of medical laboratories in the European Union. Clin Chem Lab Med 2007;45:268–75.
[4] EA 3(06): Guidelines for selection of participants to courses for the training of assessors in assessment of laboratories applying for accreditation, European co-operation of Accreditation, 1994.
[5] ILAC G11 07/2006: ILAC guidelines on qualification and competence of assessors and technical experts, ILAC, 2006.

2.19 Laboratory diagnosis of hereditary spherocytosis

May-Jean King

Summary

Hereditary spherocytosis (HS) is an inherited hemolytic anemia often diagnosed during early childhood. The appearance of spherocytes is due to a decrease in cell membrane area as a result of weakening of interactions between defective membrane protein(s) of the red cell cytoskeleton and the outer membrane lipid bilayer. The screening tests for HS are the osmotic fragility test, acid glycerol lysis time test, cryohemolysis test, ektacytometry, and the EMA binding test. Confirmation of membrane protein deficiency involves sodium dodecyl sulfate-polyacrylamide gel electrophoresis of red cell membrane proteins and quantitation by scanning densitometry. HS is a not a single gene disorder. Standard molecular biology techniques can detect protein gene mutations. Despite this, the management of HS patients is not influenced by the knowledge of protein deficiency and gene mutation(s).

Hereditary spherocytosis (HS) has a frequency of 1:2,000 to 1:5,000 births in Northern Europe. The clinical expression is heterogeneous, ranging from a silent carrier with

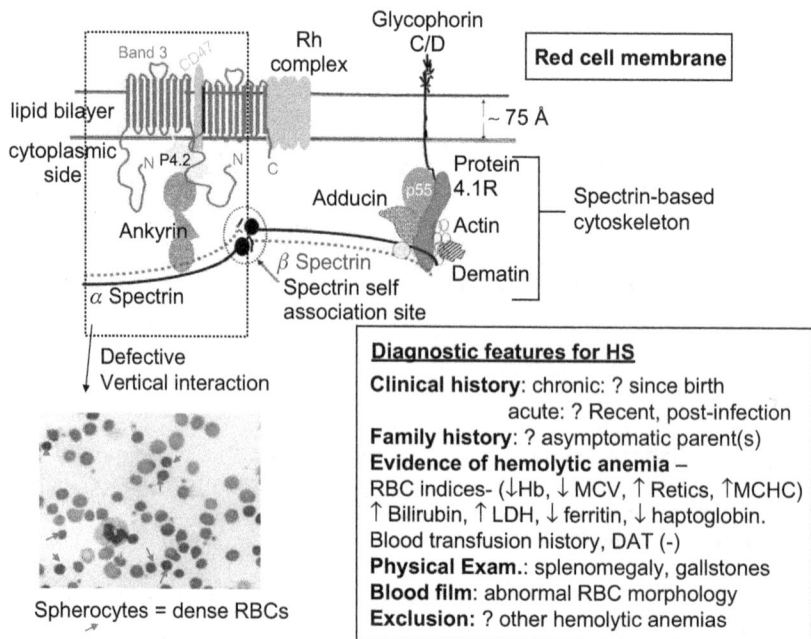

Fig. 2.19.1: Organization of red cell membrane and diagnostic features for hereditary spherocytosis (HS).

compensated hemolytic anemia to a severe hemolysis requiring blood transfusion [1]. The shortened lifespan of HS red blood cells (RBCs) in circulation is attributed to a deficiency or a functional defect in band 3 (anion transporter), protein 4.2, ankyrin, and spectrin (►Fig. 2.19.1). In normal RBCs, band 3 and glycophorin C are indirectly linked to the spectrin-based cytoskeleton via ankyrin, protein 4.2, and protein 4.1R. These cross-linkages create a three-dimensional protein network that maintains the cell shape, and confers elasticity and deformability to the RBC when traveling through blood vessels in tissues and organs. Mutation(s) in one of these proteins can result in either a protein deficiency or a loss of the ability of the defective protein to interact with its immediate neighbors in the membrane.

A rapid decline in hemoglobin levels and hyperbilirubinemia in a neonate [2] or a child with post-infection spherocytosis suggests HS. A diagnosis of HS is straightforward if the patient has a family history and presents characteristic features (►Fig. 2.19.1).

(A) Osmotic Fragility (OF) Test

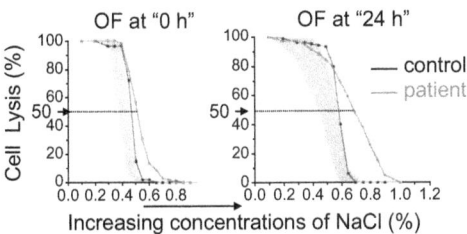

OF at "0 h" OF at "24 h"

— control
— patient

Cell Lysis (%)

Increasing concentrations of NaCl (%)

(C) Cryohemolysis Test

$$\% \text{ hemolysis} = \frac{A^{540} \text{ Test}}{A^{540} \text{ whole cell lysate}} \times 100\%$$

Results: normal = 3% –15%
HS > 20% lysis

(E) The EMA Binding Test

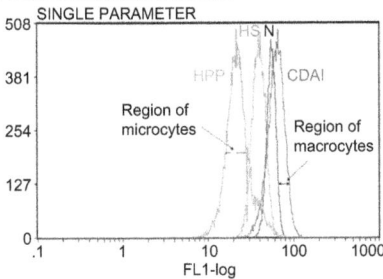

SINGLE PARAMETER

HS N
HPP
CDAI
Region of microcytes
Region of macrocytes

FL1-log

(B) Acid Glycerol Lysis Time (AGLT)

GLT_{50} = 50% hemolysis of RBC in a buffered hypotonic saline-glycerol mix
Results: normal = 30 mins.
HS = 25 s–150 s

(D) Ektacytometry

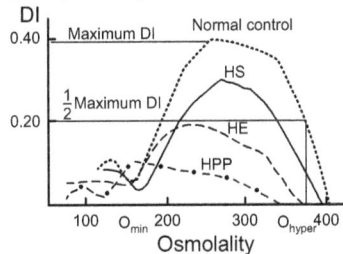

DI
Maximum DI Normal control
HS
½ Maximum DI
HE
HPP

O_{min} Osmolality O_{hyper}

(F) SDS-PAGE
(Fairbanks gel)

HS: Sp/Ank deficiency

α Spectrin
β spectrin
Ankyrin
Band 3
P4.1
P4.2

N HS N

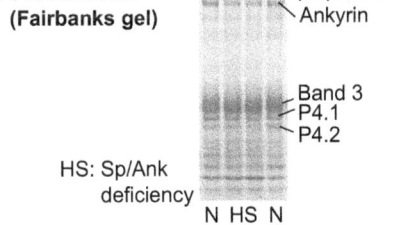

Fig. 2.19.2: Laboratory tests for hereditary spherocytosis. The results obtained from the screening tests: (A) osmotic fragility curves for RBCs incubated in increasing concentrations of NaCl. (B) acid glycerol lysis time (AGLT) results; (C) results for Cryohemolysis test, (D) deformability curves for hereditary spherocytosis (HS), hereditary elliptocytosis (HE), and hereditary pyropoikilocytosis (HPP), and (E) an overlay of histograms for HPP, HS, normal, and RBCs of congenital dyserythropoietic anemia (CDA) type I. (F) SDS-polyacrylamide gel electrophoresis of red cell membranes showing normal controls (N) and a HS sample with combined deficiency of spectrin and ankyrin.

An infection can exacerbate an extremely mild HS condition. When a patient with asymptomatic parents presents with a hemolytic anemia and spherocytosis, laboratory testing for HS is required [3].

The screening tests for detecting HS are the osmotic fragility test [4], acid glycerol lysis test (AGLT) [5], cryohemolysis test [6], ektacytometry [7] and the eosin-5'-maleimide (EMA) binding test [8] (▶Fig. 2.19.2). The osmotic fragility test, AGLT and cryohemolysis test use spectrophotometry to measure the amount of hemoglobin released from the RBCs after incubation in different solutions at 37°C, room temperature or on ice (▶Fig. 2.19.2A–C). Ektacytometry gives deformability curves for the RBCs of different disorders analyzed in mechanical stress condition (▶Fig. 2.19.2D). Flow cytometric analysis of RBCs labeled with EMA can detect (micro)spherocytes, phenotypically normal RBCs, and macrocytes (▶Fig. 2.19.2E) [9]. Sodium dodecyl sulfate-polyacrylamide gel electrophoresis of erythrocyte membrane proteins (▶Fig. 2.19.2F) and scanning densitometry are used for determining membrane protein deficiency. Sequencing of DNA or the cDNA of affected protein genes gives the genetic basis of HS within a family.

References

[1] Perrotta S, Gallagher PG, Mohandas N. Hereditary spherocytosis. Lancet 2008;372: 1411–26.
[2] Delhommeau F, Cynober T, Schischmanoff PO, Rohlich P, Delaunay J, Mohandas N, Tchernia G. Natural history of hereditary spherocytosis during the first year of life. Blood 2000;95:393–7.
[3] Bolton-Maggs PHB, Langer JC, Iolascon A, Tittensor P, King M-J. Guidelines for the diagnosis and management of hereditary spherocytosis – 2011 update. Br J Haematol 2012;156:37–49.
[4] Parpart AK, Lorenz PB, Parpart ER, Gregg JR, Chase AM. The osmotic resistance (fragility) of human red cells. J Clin Invest 1947;26:636–40.
[5] Zanella A, Izzo C, Rebulla P, Zanuso F, Perroni L, Sirchia G. Acidified glycerol lysis test: a screening test for spherocytosis. Br J Haematol 1980;45:481–6.
[6] Streichamn S, Gescheidt Y. Cryohemolysis for the detection of hereditary spherocytosis: correlation studies with osmotic fragility and autohemolysis. Am J Hematol 1998;58: 206–12.
[7] Clark MR, Mohandas N, Shohet SB. Osmotic gradient ektacytometry: comprehensive characterization of red cell volume and surface maintenance. Blood 1983;61:899–910.
[8] King M-J, Behrens J, Rogers C, Flynn C, Greenwood D, Chamber K. Rapid flow cytometric test for the diagnosis of membrane cytoskeleton-associated haemolytic anaemia. Br J Haematol 2000;111:924–33.
[9] King M-J, Telfer P, MacKinnon H, Langsbeer L, McMahon C, Darbyshire P, Dhermy D. Using the eosin-5-maleimide binding test in the differential diagnosis of hereditary spherocytosis and hereditary pyropoikilocytosis. Cytometry Part B 2008;74:244–50.

2.20 Quantification of blood folate forms using stable-isotope dilution ultra performance liquid chromatography tandem mass spectrometry

Susanne H. Kirsch, Wolfgang Herrmann and Rima Obeid

Summary

Low folate status is common and is associated with an increased risk of several diseases. Measurement of functional forms of folate can help monitor the supplement effect or study the interaction between the intake of folate and certain polymorphisms in folate enzymes. Folate levels in whole blood (WB) are less affected by recent dietary intake and they reflect long-term folate status. Few liquid chromatography tandem mass spectrometry (LC-MS/MS) methods for the determination of folate forms in WB have been described [1,2]. We describe an ultra performance liquid chromatography tandem mass spectrometry (UPLC-MS/MS) method for the quantification of 5-methyltetrahydrofolate (5-methylTHF) and non-methylfolate (sum of 5-formylTHF, 10-formylTHF, 5,10-methyleneTHF, 5,10-methenylTHF, dihydrofolate (DHF), THF, and folic acid) in WB hemolysates. The method was linear over a broad range (0.2–100 nmol/L). In WB, the limits of detection (LOD) were between 0.12 nmol/L (5-formylTHF) and 0.40 nmol/L (5,10-methenylTHF). Within day coefficients of variation CVs were between 3.4% and 9.8% for the folate forms. Recovery was between 98.8% and 101.2%. Compared to other liquid chromatography tandem mass spectrometry assays, our method is faster and has better sensitivity and selectivity.

Method

Reduced folates rapidly undergo degradation and conversion. Therefore, quantification of folate forms is challenging. We modified the recently-developed UPLC-MS/MS method for the quantification of folate forms in serum for the measurement in WB [3]. WB hemolysates were prepared by diluting ethylenediaminetetraacetic acid (EDTA) WB in 10 g/L ascorbic acid solution (pH 4.0, 0.2% Triton X-100) and incubation for 1 h at 37°C in the dark. Sample cleanup was performed by solid phase extraction (SPE) [3].

Samples were measured using an Acquity UPLC system coupled to a MicroMass Quattro Premier XE mass spectrometer (Waters Corporation, Milford, MA, USA). Folates were separated using a multi-step gradient with aqueous acetic acid at pH 2.636 and methanol over 2.5 min [3]. Identification was performed using positive electrospray ionization (ESI+) (▶Fig. 2.20.1).

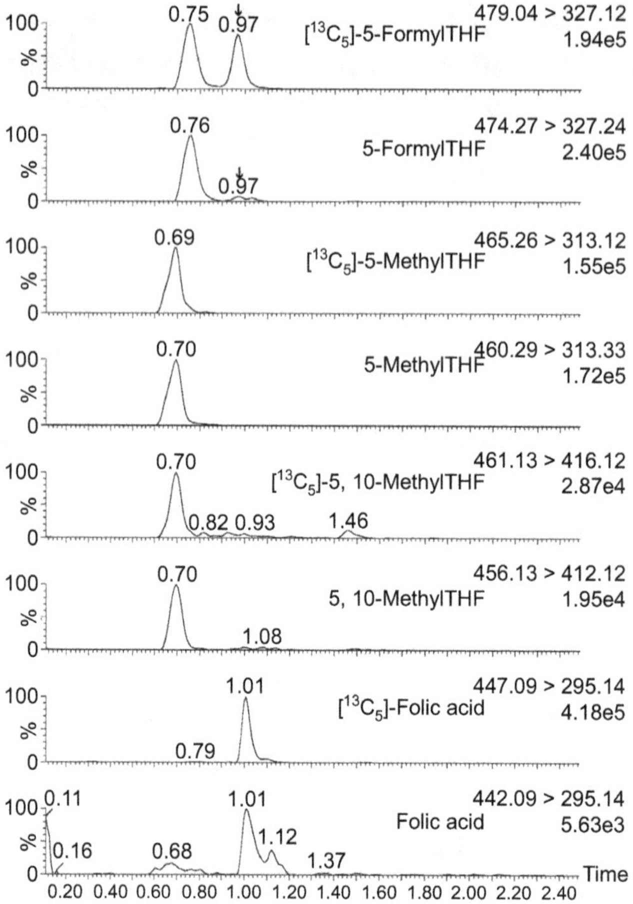

Fig. 2.20.1: Chromatogram of different folate forms in a whole blood (WB) sample. *m/z* transitions and peak intensities are shown in the upper right.

Results and discussion

The results are shown as 5-methylTHF and non-methylTHF (5-formylTHF, 10-formylTHF, 5,10-methyleneTHF, 5,10-methenylTHF, DHF, THF, and folic acid). This was due to the conversions of the folate forms during sample preparation and measurement. The inter-conversions could not been prevented by the addition of antioxidants. The method was linear between 0.2 and 100 nmol/L. The LOD for 5-methylTHF was 0.17 nmol/L in WB samples with concentrations near the instrumental detection limit. Within-day CV for 5-methylTHF was 3.4% and mean recovery was 98.8%.

Folate forms were measured in the WB of 48 non-supplemented people at baseline and after 6 months of receiving either a placebo or a vitamin supplement containing 500 μg folic acid, 50 mg B_6, and 500 μg B_{12} per day (►Tab. 2.20.1). Fasting blood was

Tab. 2.20.1: Median serum and whole blood (WB) concentrations (nmol/L) of folate forms in 48 subjects.

	Baseline		After treatment	
	Placebo group (n = 23)	Vitamin group (n = 25)*	Placebo group (n = 23)	Vitamin group (n = 25)
Serum				
5-MethylTHF	18.8	16.1	16.8	52.0[‡]
5-FormylTHF	0.28	0.19	0.44	0.38
5,10-MethenylTHF	0.14	0.15	0.08	0.48[‡]
THF	1.68	1.11	1.51	2.28[†]
Folic acid	0.00	0.00	0.10	0.11
WB				
5-MethylTHF	455	445	545	1,237[‡]
Non-methylTHF	65.8	71.2	84.8	156[‡]

*p values > 0.05 (Mann-Whitney U test). [‡]$p < 0.001$; [†]$p = 0.006$ compared to the placebo after treatment.

collected. The study was approved by the local Ethics Commission and all participants signed informed consents.

Neither group showed significant differences in folate form concentrations at baseline (▶Tab. 2.20.1). 5-MethylTHF was the predominant folate form in serum and WB (polyglutamates) and baseline total folate was in the normal range for a country without fortification. Folic acid supplementation caused a significant increase in all folate forms (except 5-formylTHF and folic acid) in serum and WB.

Conclusion

We developed a sensitive reliable high-throughput UPLC-MS/MS method for the quantification of 5-methylTHF and non-methylTHF (the sum of 5-formylTHF, 10-formylTHF, 5,10-methyleneTHF, 5,10-methenylTHF, DHF, THF, and folic acid) in WB. The method is linear between 0.2 and 100 nmol/L ($r^2 \geq 0.999$). In WB, LODs were 0.17 nmol/L for 5-methylTHF, 0.12 nmol/L for 5-formylTHF, 0.40 nmol/L for 5,10-methenylTHF, and 0.15 nmol/L for folic acid. Within day CV percentages were between 3.4% and 9.8%, with a mean recovery between 98.8% and 101.2% for the different forms.

Compared to earlier LC-MS/MS procedures, our UPLC-MS/MS method has better sensitivity and selectivity for the quantification of folate forms. The relatively short time for sample preparation (40 samples in 180 min) and measurement (2.5 min/sample) allows for the use in large-scale clinical studies.

We could confirm that 5-methylTHF was the predominant folate form in serum and WB (polyglutamates) of 48 non-supplemented and non-fortified adults. Baseline total folate in serum and WB was in the normal range for a country without fortification. Supplementation with folic acid, B_6, and B_{12} caused a significant increase in all folate forms (except 5-formylTHF and folic acid) in serum and WB.

References

[1] Fazili Z, Pfeiffer CM, Zhang M, Jain R. Erythrocyte folate extraction and quantitative determination by liquid chromatography-tandem mass spectrometry: comparison of results with microbiologic assay. Clin Chem 2005;51:2318–25.

[2] Smith DE, Kok RM, Teerlink T, Jakobs C, Smulders YM. Quantitative determination of erythrocyte folate vitamer distribution by liquid chromatography-tandem mass spectrometry. Clin Chem Lab Med 2006;44:450–9.

[3] Kirsch SH, Knapp JP, Herrmann W, Obeid R. Quantification of key folate forms in serum using stable-isotope dilution ultra performance liquid chromatography-tandem mass spectrometry. J Chromatogr B Analyt Technol Biomed Life Sci 2010;878:68–75.

2.21 Evaluation of the new Marburg cerebrospinal fluid model with human spondylopathies

*Tilmann Otto Kleine, Christa Löwer, Siegfried Bien,
Reinhardt Lehmitz and Alexandra Dorn-Beineke*

Summary

The Marburg model is based on the theorem that a few proteins (<200 mg/L) filtered from blood through the molecular sieve of blood-cerebrospinal fluid barrier (B-CSF-B) have blood proteins added to them in the lymph, traveling from paravertebral lymph vessels into the spinal CSF; thus proteins increase to <420 mg/L. Normally, CSF flows out from spinal canal into paraspinal lymph vessels. The upright human position reverses the equilibrium between CSF and lymph pressures, allowing minimal lymph leakage into the lumbar CSF: 0.3 mL of lymph elevates CSF proteins from 200 to 400 mg/L. Prolapses alter CSF backbone clearance via a valve mechanism, the disc prolapse bulges into the spinal cord and increases the resistance of CSF outflow. Lymph admittance is facilitated, however, thus more lymph enters and total amount of protein increases in the lumbar CSF. Molar IgG/albumin ratios show the entrance of plasma proteins in cases of lumbar prolapses. The IgG/albumin ratio increases from 0.049 nmol/nmol with controls up to 0.065 nmol/nmol in individuals with mass prolapses in lumbar CSF; thus approximating an IgG/albumin ratio of 0.125 in normal human plasma.

The blood-brain barrier and blood-cerebrospinal fluid barrier

Human central nervous system (CNS) is separated from the body by different barriers [1]. The blood-brain barrier (BBB) is localized in the capillaries of the brain and spinal cord. Endothelia cells are completely sealed by a complex of 2 tight junctions; therefore, no blood proteins pass the BBB into the cerebrospinal fluid (CSF).

The blood-CSF barrier is localized in a complex of 1 tight junction between the epithelia of the plexus choroidei in brain ventricles. It is a molecular sieve that filters blood proteins into the CSF in the molar ratio of albumin:IgG:IgA:IgM of 28,000:1,600:50:1; revealing an IgG:albumin ratio of 0.057 nmol/nmol.

Total protein increases to 200 mg/L in the suboccipital (SOP) CSF. This is measured with the biuret reaction of total protein precipitated with trichloroacetic acid (CV <8%, inaccuracy <10%); albumin, IgG, IgA and IgM with modified immune nephelometry (CV <9%, inaccuracy <10%) [2].

CSF resorption

CSF is clarified 3–5 times per day [3]. About half of CSF flow being intra-cranial, half of the CSF penetrates along the nerve roots into valveless lymph vessels to flow into the paraspinal lymphatic system. There is an unstable equilibrium between the pressure of the intraspinal CSF column and the pressure in the paraspinal lymph vessels. This is modulated by pressing, sneezing or coughing. Small quantities of lymph, which are rich in protein, raise CSF proteins in the suboccipital > cervical > lumbar CSF. Here, 0.3 mL lymph elevates the total protein level from 200 mg/L in SOP to 400 mg/L in lumbar CSF! Lymph (Yml), added to SOP CSF, is calculated with polynomial formulae considering $X = \sum$ pMole albumin, IgG in normal SOP CSF plus pMole albumin, IgG per mL of normal human plasma. The bilateral validity of 4 formulae is revealed.

CSF resorption with prolapses

Disk prolapses penetrate into the spinal canal and compresses the spinal cord [4]. Thus CSF flow is locally impaired; but not lymph flow into the spinal canal. Lymph proteins enter and increase the total protein in the lumbar CSF (►Fig. 2.21.1). Lumbar Solutrast® myelography supplies the relevant radiological diagnosis. Our findings contrast with earlier data of lower back pain syndromes reporting 81% of protein values to be within the normal range [5].

The valve mechanism of lumbar prolapse explains the diminished CSF flow by increased local resistance simultaneously with the increased entrance of lymph by facilitated lymph in-flow: 0.31 mL lymph enters lumbar CSF with protrusion (small disk prominence); 0.37 mL with lumbar prolapse (middle disc bulges); and 0.51 mL lymph with mass prolapse (most of middle disc bulges into spinal cord) (►Fig. 2.21.1).

Evidence of plasma proteins entering the lumbar CSF gives molar IgG:albumin ratios, since no intrathecal Ig production is revealed with normal index IgG, IgA and IgM and negative oligoclonal bands in all CSF samples analyzed. With controls, the mean IgG: albumin ratio is calculated to be 0.049 nmol/nmol in lumbar CSF. The IgG:albumin ratio increases with the dimensions of prolapse penetration up to 0.065 nmol/nmol (►Fig. 2.21.1). Thus the IgG:albumin molar ratio approximates to 0.126 of normal human serum.

Conclusion

The Marburg CSF model explains the origin of proteins in the CSF. The BBB in CNS capillaries is tight to blood proteins. The blood-CSF barrier filters blood proteins through molecular sieve to 200 mg/L protein. The occurrence of 0.3 mL lymph diffusing from paravertebral lymph vessels into the lumbar CSF increases protein to 420 mg/L with blood proteins.

The valve mechanism of lumbar prolapse facilitates the entrance of lymph. More blood proteins (in lymph) are added into the lumbar CSF to proteins filtered through blood-CSF barrier, thus increasing proteins to 754 mg/L with mass prolapses (►Fig. 2.21.1).

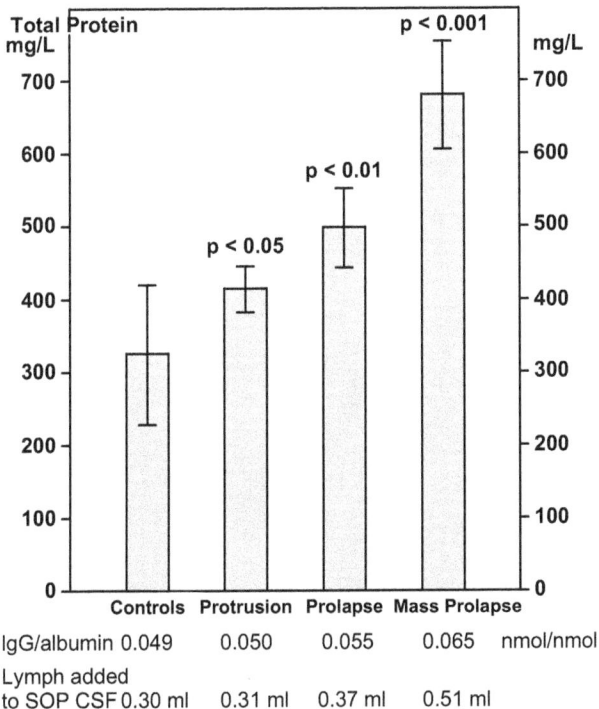

Fig. 2.21.1: Total protein (mg/L mean values ± standard deviation) in the lumbar cerebrospinal fluid of control [n = 145, specified by [2], protrusion (n = 45), lumbar prolapse (n = 110), and mass prolapse (n = 18) patients.
Significant mean differences to controls calculated with t-test ($p < 0.05$). Molar IgG/albumin ratio is significantly higher with lumbar and mass prolapses.

Three CSF types can be found: molecular-sieve CSF, spinal CSF with lymph proteins, and blood proteins-enriched CSF with inflammatory-destroyed BBB [TO Kleine, personal communication]. Our findings contrast with the statements that there are "low selective barriers" operating with barrier dysfunction [6,7] where all of the proteins pass the blood-CSF barrier by passive transport [8].

References

[1] Kleine TO. Blut-Hirn-Schranke-Funktionsteste. In: Gressner AM, Arndt T, eds. Lexikon der Medizinischen Laboratoriumsdiagnostik. Heidelberg: Springer Medizin Verlag; 2007, pp. 206–11.
[2] Kleine TO, ed. Neue Labormethoden für die Liquordiagnostik. Stuttgart, New York: Thieme; 1980.
[3] Kleine TO. Nervensysytem. In: Greiling H, Gressner AM, eds. Lehrbuch der Klinischen Chemie und Pathobiochemie. Stuttgart, New York: Schattauer Verlag; 1989, pp. 860–93.
[4] Batynski WS, Ortiz AO. Interventional assessment of the lumbar disk: provocation lumbar discography and functional anesthetic discography. Tech Vasc Interv Radiol 2009;12:33–43.

[5] Ahonen A, Myllylä W, Hokkanan E. Cerebrospinal fluid findings in various lower back pain syndromes. Acta Neurol Scand 1979;60:93–9.
[6] Felgenhauer K. Protein filtration and secretion at human body fluid barriers. Pflügers Arch 1980;384:9–17.
[7] Reiber H, Felgenhauer K. Protein transfer at the blood cerebrospinal fluid barrier and the quantitation of the humoral immune response within the central nervous system. Clinica Chimica Acta 1987;163:319–28.
[8] Felgenhauer K. Protein size and cerebrospinal fluid composition. Klin Wschr 1984;52:1158–64.

2.22 E-learning experiences of national societies of clinical chemistry and laboratory medicine

Petr Kocna

Summary

E-learning and distance education opportunities in clinical chemistry were reviewed in 2005 by the International Federation of Clinical Chemistry (IFCC) Working Group on Distance Education chaired by D. Juretic, under the auspices of the Committee on Education and Curriculum Development chaired by L.C. Allen. A second survey of national societies was carried out in 2010 by the Committee on Education and Curriculum Development chaired by P. Kocna and the Working Group on Internet-Distance Learning chaired by V.T. Thanh. The aim of this study was to summarize changes and trends in the use of e-learning and distance education by member societies of the IFCC over a 5-year period. Surveys of national societies were performed using printed questionnaires in 2005 and by on-line questionnaire forms in 2010. The response rate from member-societies was ~50% (34 in 2005 and 42 in 2010). National society websites increased from 70.6% in 2005 to 90.5% in 2010. National society websites with educational sections increased from 41.2% in 2005 to 57.1% in 2010. Lectures and presentations published on websites are still the most widely used form of educational resource (79.2% in 2010 and 71.4% in 2005). In 2008 the IFCC Committee on Education and Curriculum Development published an on-line Educational Resource Database recommending 254 resources for education in clinical chemistry and was visited more than 1790 times by visitors from 86 countries. This database is regularly used by 12 national societies and 40% of societies recommended distance education on their websites. The internet educational resource most often recommended by national society websites was the NLM PubMed database (mean mark 2.3) followed by Google (mean mark 2.38). A majority of national societies (76.2%) preferred a unified IFCC educational strategy and many responses promoted the concept of IFCC education credits (59.5% of responding national societies).

The use of e-learning education in clinical chemistry

The International Federation of Clinical Chemistry (IFCC) Committee on Education and Curriculum Development (CECD) and Working Group on Distance Education (WGDE jointly drafted the first IFCC survey on website education, e-learning and distance education in May 2004 while at a meeting in Sousse, Tunisia. This first survey was conducted with printed questionnaires circulated to national societies who were members by mail and was analyzed in January 2005 (by D. Juretic, chair of WGDE and L.C. Allen chair CECD). A second survey of national societies was carried out in 2010 via an on-line internet questionnaire (produced by P. Kocna, chair of the CECD and V.T. Thanh, chair of the Working Group on Internet-Distance Learning).

This study summarizes changes and trends in the use of e-learning and distance education in clinical chemistry over the 5-year period between 2005 and 2010. Responses were received from approximately 50% of member societies for the two surveys (34 in 2005 and 42 in 2010). National society websites increased from 70.6% (24 of 34) in 2005 to 90.5% (38 of 42) in 2010, and the educational sections of websites increased from 41.2% (14 of 34) in 2005 to 57.1% (24 of 42) in 2010. The most important changes were 5-fold increases in undergraduate study programs, e-mail teaching communications and the use of electronic books (▶Tab. 2.22.1).

The responses from 2005 to 2010 listed in ▶Tab. 2.22.1 show a dramatic increase in e-learning activities in clinical chemistry. This trend affected most medical disciplines and was not unique to clinical chemistry. This trend was fostered in Europe and the Czech Republic in particular in 2006 when the network of medical faculties MEFANET (MEdical FAculties NETwork [1]) was created. The MEFANET aimed to efficiently share medical teaching materials. It achieved its goals using web portals that were accessible to all member faculties using a common internet gateway. Medical teaching materials were shared among 10 faculties of the Czech and Slovak Republics. This unique collaborative environment allowed educational resources to be easily shared and in turn this led to the rapid adoption and growth of the network.

In 2008 the IFCC CECD published an on-line Educational Resource Database recommending 254 resources for education in clinical chemistry (http://eduweb.virt. cz). The database has been visited thousands of times by visitors from 86 countries. The database was developed on Common Gateway Interface (CGI) scripts – EZDB.CGI programmed in Perl by S. Barde, running on the 1st Medical Faculty of Charles University web-server. The database environment has been used in the past for other educational resources, electronic journals and medical images [2]. The database attempts to cover all main clinical chemistry topics [3,4], as published in *Selected Modern Curricula and Training Program Requirements for Training in Clinical Chemistry and Clinical Laboratory Medicine*, where web-based educational resources are available [http://www1.lf1.cuni.cz/~kocna/edu/].

The second survey of national societies determined the preferences for internet-based educational resources used in different countries. The internet educational resource

Tab. 2.22.1: Percentage of national society websites with specific educational materials based on survey responses in 2005 and 2010.

Web-based education activity	2005	2010
Undergraduate study program in CBLM	2.9%	11.9%
Postgraduate study program in CBLM	20.5%	38.1%
Program of specialization in CBLM	14.7%	28.5%
Clinical cases	5.8%	11.9%
Multiple choice questions	2.9%	7.1%
E-mail teaching communication	2.9%	16.7%
Electronic books	2.9%	16.7%
Educational use of EQA national programs	17.6%	9.5%
Lectures of congresses or courses	29.4%	45.2%

CBLM, Clinical Biochemistry and Laboratory Medicine; EQA, external quality assessment.

most often recommended by national societies was the NLM PubMed database (mean mark 2.3) followed by the Google search engine (mean mark 2.38). Most national societies (76.2% – 32 out of 42) preferred a unified IFCC educational strategy promoted by the concept of IFCC educational credits (59.5% – 15 out of 42 societies).

References

[1] Stipek S, Dusek L, Schwarz D, Stuka C, Vejrazka M, Nikl T. MEFANET – a new kind of network for electronic support of medical and health care education. In: Siemens G, Fulford C., eds. Proceedings of the world conference on educational multimedia, hypermedia and telecommunications. Chesapeake, VA: AACE; 2009, pp. 1323–5.
[2] Dohnal L, Kocna P. Electronic Journal – clin biochem on the web using a database CGI script. Clin Chem Lab Med 2001;39:S285.
[3] Allen LC. Clinical case material for teaching clinical chemistry and laboratory medicine. Clin Chem Lab Med 2001;39:875–89.
[4] Fink NE, Allen LC. Handbook on master program in clinical laboratory sciences. Clin Chem Lab Med 2003;41:1379–86.

2.23 The experience of Médecins Sans Frontières in laboratory medicine in resource-limited settings

Cara S. Kosack

Summary

In medical humanitarian assistance, the diagnosis of diseases plays a crucial role in Médecins Sans Frontières (MSF) programs. Laboratory investigations are one of the main diagnostic tools utilized in these programs. Currently MSF supports and/or operates more than 130 laboratories in approximately 45 countries. The variety of analysis offered depends largely on the context of the program and the availability of context-adapted tools and ranges from sophisticated laboratories specializing in tuberculosis culture to small laboratories within a primary healthcare program or operating as mobile clinics. Within MSF as a whole, 11 laboratory advisors currently support MSF field laboratories and staff and provide expertise and advice on relevant diagnostic laboratory techniques, tools, and testing guidelines and policies.

Introduction

Depending on the type of field medical program supported by Médecins Sans Frontières (MSF), different laboratory analysis and tests are provided as a minimum standard (▶Tab. 2.23.1). The largest laboratories in MSF are found in vertical tuberculosis (TB) and HIV programs. Other MSF programs are either disease-specific (e.g., malaria, Chagas, kala azar [visceral leishmaniasis], sleeping sickness, malaria, malnutrition, and sexually transmitted infections) or are integrated as part of primary or secondary healthcare structures.

Case study: sleeping sickness

The complexity of diagnostic algorithms and protocols faced in the field by MSF is exemplified by the testing needs for sleeping sickness (human African trypanosomiasis, or HAT; ▶Fig. 2.23.1). For the diagnosis of HAT, screening is carried out by using the card agglutination test for trypanosomiasis, followed by confirmation microscopically tests detecting the *Trypanosoma* parasites. Confirmation methods are parasite detection in lymph node aspirates and isolation of parasites from whole blood using capillary tube centrifugation technique and/or mini-anion exchange centrifugation technique (mAECT).

For determining HAT disease stage, laboratory investigations are performed on the cerebrospinal fluid of patients. White blood cells (per mm^3) are counted and *Trypanosoma* are detected following centrifugation. In areas of high HAT prevalence (>1%) where mAECT is not available, treatment can be started on the basis of a card agglutination test for trypanosomiasis-positive result in a 1:16 dilution (strong serological suspect), without parasitological confirmation. Disease staging remains necessary for selecting

Tab. 2.23.1: Overview of laboratory standard analysis by program type.

Program type	Laboratory analysis
Tuberculosis	Microscopy: Ziehl-Neelsen or Auramine, Xpert: MTB/RIF Culture: MGIT or thin layer agar (TLA)
HIV	Testing: via lateral flow immunoassays e.g. Determine, Uni-Gold Staging: CD4 count and CD4% Monitoring HAART: lactate, hemoglobin, ALAT, ASAT, creatinine
Malaria	Microscopy and/or lateral flow immunoassays, Potentially blood transfusion
Blood transfusion	Blood grouping and cross matching Infectious disease screening: HIV, hepatitis B and C, syphilis
Chagas disease	Lateral flow immunoassays Indirect hemagglutination (IHA) ELISA
Sleeping sickness (human African trypanosomiasis)	CATT Microscopy lymph node aspirates, mAECT, WOO test Disease staging: cerebrospinal fluid
Kala azar (visceral leishmaniasis)	Lateral flow immunoassay: rK 39 Direct agglutination test (DAT) Microscopy of lymph node and spleen aspirates
Clinical chemistry	Wet and dry chemistry
Hematology	Microscopy and chamber counts as well as small hematology analyzers
Sexual and reproductive health	Syphilis testing via lateral flow immunoassays, urine dipsticks and microscopy
Others	Brucellosis, dengue, Gram-staining in bacterial infections, meningitis, cholera

MTB/RIF, *Mycobacterium tuberculosis*/rifampicin; MGIT, mycobacteria growth indicator tube; TLA, thin layer agar; CD4, T4 lymphocyte; HAART, highly active antiretroviral therapy; ALAT, alanine aminotransferase; ASAT, aspartate aminotransferase; HIV, human immunodeficiency virus; IHA, indirect hemagglutination; ELISA, enzyme-linked immunosorbent assay; CATT, card agglutination test for trypanosomiasis; mAECT, mini anion exchange centrifugation technique; WOO, named after scientist Woo; DAT, direct agglutination test.

the appropriate treatment. Besides the complexity of HAT diagnosis and the urgent need for simplified diagnostic tools, one of the main problems regarding HAT diagnosis is access to an uninterrupted supply of diagnostic materials [1,2].

Clinical chemistry in MSF

Clinical chemistry is mostly used when monitoring treatment, for example in HIV or TB treatment programs. For the most part "wet" chemistry analyzers such as the HumaLyzer 2000 or HumaLyzer Junior are used. Programs with low demand prefer opting for "dry"

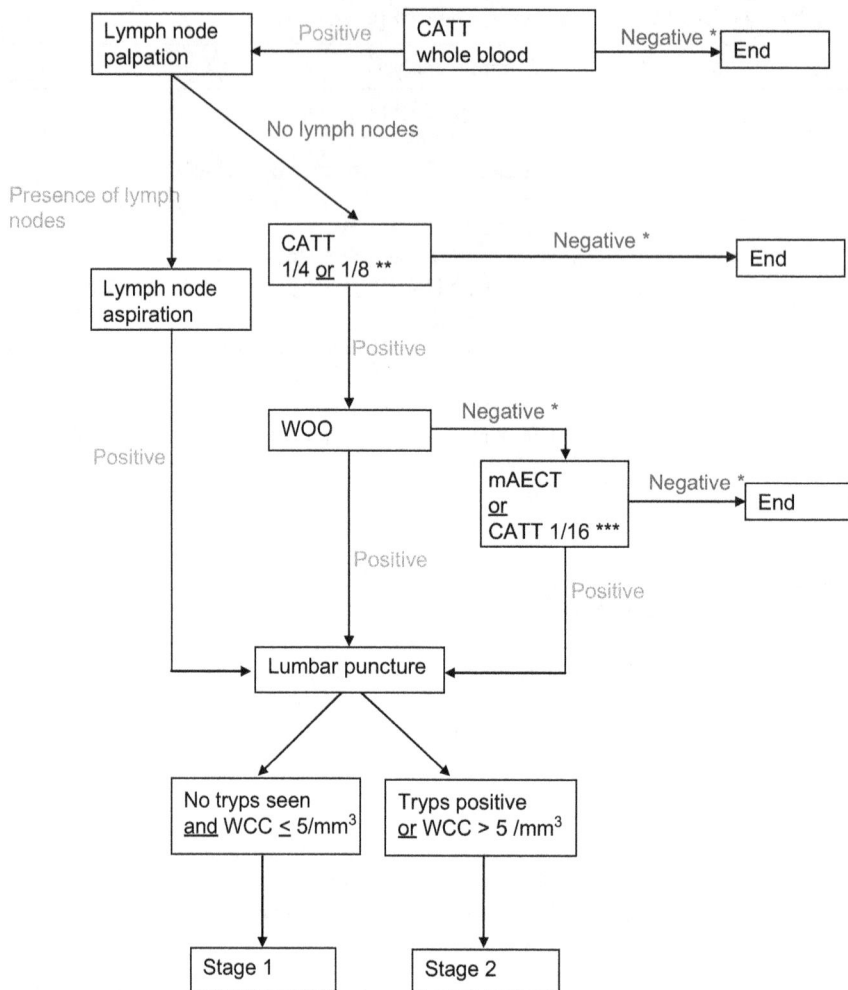

Fig. 2.23.1: Médecins Sans Frontières diagnostic algorithm for sleeping sickness [human African trypanosomiasis (HAT)].
*If high clinical suspicion of HAT, move to next step of the diagnostic tree.
**If prevalence <1% use cut-off of card agglutination test for trypanosomiasis (CATT) 1/8.
***Use CATT 1/16 if mini anion exchange centrifugation technique (mAECT) not available and prevalence >1%.

chemistry due to the maintenance and quality control needs of the analyzer, but the costs of using dry chemistry are high (€1–2 per test compared with a few cents when using wet methods). In MSF laboratories, the number of tests ordered for wet chemistry analyzers is much higher than for dry analyzers (▶Fig. 2.23.2).

In 2010, MSF contemplated moving away from wet chemistry systems towards using dry chemistry systems, such as the Reflotron® by Roche. A cost-calculation model showed that using only dry chemistry would mean an increase in cost of more than 50 times, which is not financially sustainable.

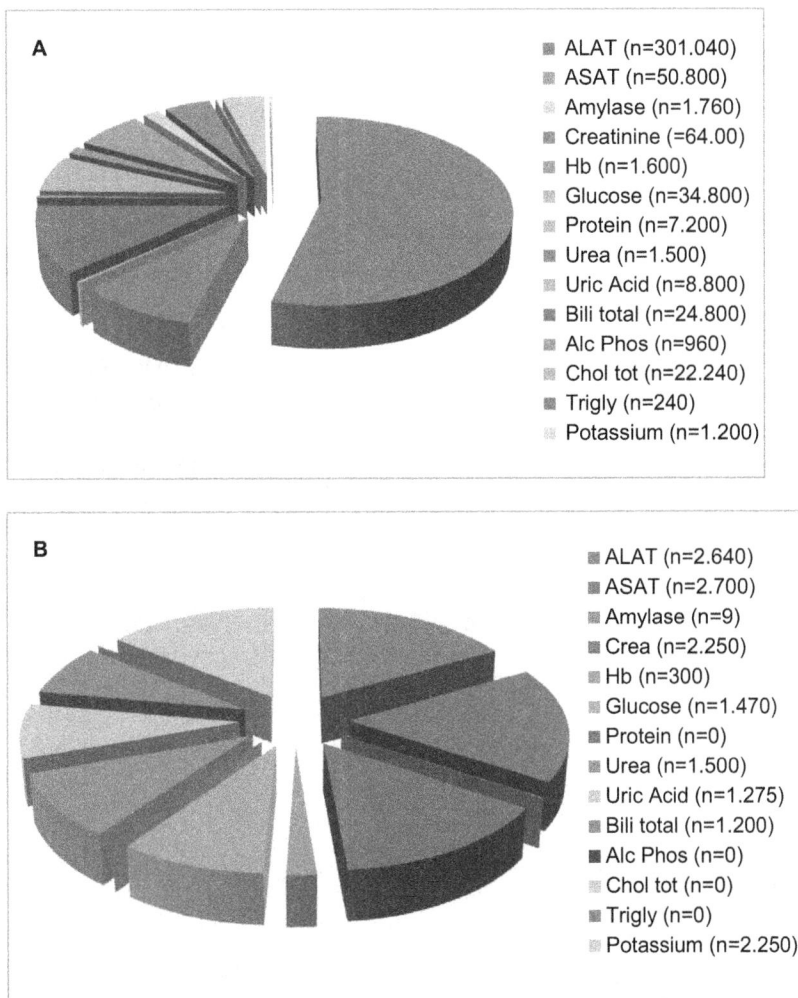

Fig. 2.23.2: (A) Numbers of "wet" clinical chemistry tests sent to Médecins Sans Frontières field programs in 2009.
(B) Numbers of "dry" clinical chemistry tests sent to Médecins Sans Frontières field programs in 2009.

Laboratory quality control

MSF considers quality control for laboratory testing to be a high-priority and has developed a quality control protocol for microscopy that is more applicable to resource-limited settings than other approaches, such as lot quality assurance sampling. The MSF quality control approach was designed to [3]:

- allow a small sample size to be feasible across all settings;
- enable reliable analysis;
- monitor both false-positive and false-negative results; and
- be applicable to all microscopy testing.

In 2010, MSF enrolled several laboratories in the sub-Saharan region in the international accredited proficiency testing provided by the National Health Laboratory Services and National Institute of Communicable Diseases in South Africa. Over time, external quality control has allowed improvement of performance in the different areas, such as malaria, TB microscopy, biochemistry, hematology and CD4 testing.

New technologies

MSF aims to advocate for access to, and simplification of, diagnostic tools in resource-limited settings and monitors market developments on an ongoing basis. One recent example is provided by the Xpert MTB/Rif assay for the diagnosis of TB and rifampicin resistance and endorsed by the World Health Organization (WHO) in December 2010 as the initial diagnostic test in settings characterized by a high prevalence of multi-drug resistant-TB and HIV.

The Xpert system allows a rapid TB diagnosis (turn-around-time <2 h) and includes testing for resistance to rifampicin. Within one cartridge the test integrates and automates the three processes required for real-time PCR-based molecular testing of TB: sample preparation, amplification, and detection. Studies have demonstrated good performance of this test on pulmonary samples, with a sensitivity and specificity approaching that of culture, which is considered the gold standard for TB diagnosis [4]. The system also shows favorable results for the diagnosis of some of forms of extrapulmonary TB [5].

MSF participated in the multicentric demonstration study of this new technology in Khayelitsha, South Africa, in collaboration with the University of Cape Town and the Foundation for Innovative New Diagnostics [6]. It will implement more than 20 analyzers in its programs in the future.

Bottle necks

The WHO's ASSURED criteria outline the ideal characteristics for the design of a diagnostic test for resource-limited settings as Affordable, Sensitive, Specific, User-friendly, Rapid and robust, Equipment-free, and Deliverable to those in need [7]. Although originally developed by the WHO's Sexually Transmitted Diseases Diagnostics Initiative, the ASSURED criteria have been used for other diseases as well, summarizing useful criteria for effective diagnostic tools for diseases found in resource-limited settings.

Conclusions

Access to reliable diagnostic tools in humanitarian settings is a cornerstone of quality medical humanitarian assistance. MSF's experience in reinforcing laboratory capacity over the past decade has shown that much can be done to improve diagnostic capacity, even in the most remote and unstable settings. However, the extent to which improvements can be made depends to a considerable degree on the extent to which diagnostic tools are developed with humanitarian settings in mind. Such settings often lack skilled health staff, have limited health budgets, are confronted with a high burden of infectious diseases with most patients accessing care at the primary care level, and are far from secondary or tertiary hospitals. Unless efforts are made to ensure that advances in diagnostic technology take into account the realities of resource-limited settings, they are likely to be irrelevant to the majority of patients who could benefit from it.

Acknowledgements

The author wishes to thank the MSF Laboratory Working Group. Without their continuous support and dedication, no advance in laboratory diagnosis in MSF would have been possible. Special thanks are extended to Nathan Ford, Oliver Yun, Martina Casenghi, Teri Roberts, Emmanuel Fajardo, Saskia Spijker and Roberto de la Tour for valuable comments and effective assistance during manuscript preparation.

References

[1] Stich A, Barrett MP, Krishna S. Waking up to sleeping sickness. Trends Parasitol 2003;19:195–7.
[2] Buescher P, Ngoyi DM, Kabore J, Lejon V, Robays J, Jamonneau V, Bebronne N, Van der Veken W, Bieler S. Improved models of Mini Anion Exchange Centrifugation Technique (mAECT) and Modified Single Centrifugation (MSC) for sleeping sickness diagnosis and staging. PLoS Negl Trop Dis 2009;3:e471.
[3] Klarkowski DB, Orozco DJ. Microscopy quality control in Médecins Sans Frontières programs in resource-limited settings PLoS Med 2011;7(1):e1000206.
[4] Boehme CC, Nabeta P, Hillemann D, Mark NP, Shenai S, Krapp F, Allen J, Tahirli R, Blakemore R, Rustomjee R, Milovic A, Jones M, O'Brien SM, Persing DH, Ruesch-Gerdes S, Gotuzzo E, Rodrigues C, Alland D, Perkins MD. Rapid molecular detection of tuberculosis and rifampicin resistance. N Engl J Med 2010;363:1005–15.
[5] Hillemann D, Ruesch-Gerdes S, Boehme C, Richter E. Rapid molecular detection of extrapulmonary tuberculosis by automated GeneXpert® MTB/RIF system. J Clin Microbiol 2011;49:1202–5.
[6] Boehme CC, Nicol MP, Nabeta P, Michael JS, Gotuzzo E, Tahirli R, Gler MT, Blakemore R, Worodria W, Gray C, Huang L, Caceres T, Mehdiyev R, Raymond L, Whitelaw A, Sagadevan K, Alexander H, Albert H, Cobelens F, Cox H, Alland D, Perkins MD. Feasibility, diagnostic accuracy, and effectiveness of decentralised use of the Xpert MTB/RIF test for diagnosis of tuberculosis and multidrug resistance: a multicentre implementation study. Lancet 2011;377:1495–505.
[7] WHO/TDR. Diagnostics for tuberculosis: global demand and market potential. 2006.

2.24 Screening, identifying, and quantifying small molecules by hyphenated mass spectrometry in toxicology and drug monitoring – an update

Hans H. Maurer

Summary

Background: Reliable analytical data are important prerequisites for competent assessments in clinical and forensic toxicology as well as clinical pharmacology. The analytical strategy may include screening, confirmation and identification (in toxicology) followed by quantification of relevant compounds and pharmacokinetics-based interpretation of the results.

Methods: Mass spectrometry coupled to gas chromatography (GC-MS) or liquid chromatography (LC-MS) is the gold standard in clinical and forensic toxicology because of its universality, reliability, high-sensitivity and specificity. For the same reasons, LC-MS has also become the gold standard in drug monitoring.

Results: GC-MS and increasingly LC-MS are used for target and comprehensive screening, library-assisted identification, and validated quantification of drugs, poisons and their metabolites in blood, urine or alternative matrices. Concepts and procedures using GC-MS or LC-MS techniques in the areas of toxicology and drug monitoring with a special focus on multi-analyte procedures are presented and discussed.

Conclusions: The reliability of GC-MS and LC-MS helps to ensure the quality of analytical data needed for the correct interpretation of analytical findings, thus helping to avoid incorrect treatment of the patient or analytical data being contested in court.

Introduction

An efficient toxicological analysis is the basis of competent toxicological judgment, consultation and expertise. In clinical and forensic toxicology, the drug or poison of interest must often be screened for and identified before quantification. Low or high-resolution mass spectrometry (MS) coupled to gas chromatography (GC) or liquid chromatography (LC) are the methods of choice due to their high-selectivity, sensitivity and universality [1,2]. In recent years, LC-MS has become increasingly popular since most of its former drawbacks have been overcome. In the following, typical examples for screening, identification and quantification procedures are critically reviewed.

Screening procedures

Several concepts have been developed for multi-target screening in urine and blood using the selected-ion monitoring or multiple-reaction monitoring mode [1,2]. Special applications were published for urine testing by dilute and shoot [1,2], for breath testing by exhale and shoot [3], or for dried blood spots by desorb and shoot [4]. For non-target, comprehensive screening procedures, library-based approaches were described. Besides this, a series of GC-MS papers on screening for emerging drugs of abuse using huge reference spectra collections [5], LC-MS approaches using high-resolution single stage time-of flight (TOF) or tandem time-of flight analyzers (QTOF) were published [6,7]. A new metabolite-based LC-MS^n (LC-MS to the stage of n) urine screening was developed using linear ion trap technology allowing us to use the full spectral information of the MS^2 and MS^3 spectra [8]. The transferability of the new library consisting of over 2,300 metabolite and 1,000 drug spectra to a QTRAP mass analyzer was studied successfully using the sophisticated software tool SmileMS [9].

Quantification of drugs and their metabolites

Multi-analyte procedures for quantification of drugs and/or their metabolites in body samples have recently been critically reviewed in detail [1,2]. A new multi-analyte approach for the determination of over 100 drugs from 6 different drug classes was developed, intensively tested for ion suppression or enhancement caused by the various deuterated internal standards and overlapping analytes, and finally fully validated [10].

Conclusion

GC and LC coupled to low or high-resolution MS are indispensable tools in clinical and forensic toxicology. They provide the high-selectivity, sensitivity and universality needed for screening, identification and quantification of drugs, poisons and/or their metabolites in body samples. While LC-MS/MS is the method of choice for target screening and quantification, low- and high-resolution linear ion traps are the most powerful techniques for more comprehensive screenings. High-resolution can also be useful if supplementary fragmentation increases the identification power.

References

[1] Maurer HH. Perspectives of liquid chromatography coupled to low and high resolution mass spectrometry for screening, identification and quantification of drugs in clinical and forensic toxicology [review]. Ther Drug Monit 2010;32:324–7.

[2] Peters FT. Recent advances of liquid chromatography-(tandem) mass spectrometry in clinical and forensic toxicology [review]. Clin Biochem 2010;44:54–65.

[3] Beck O, Sandqvist S, Eriksen P, Franck J, Palmskog G. Determination of methadone in exhaled breath condensate by liquid chromatography-tandem mass spectrometry. J Anal Toxicol 2011;35:129–33.

[4] Thomas A, Deglon J, Steimer T, Mangin P, Daali Y, Staub C. On-line desorption of dried blood spots coupled to hydrophilic interaction/reversed-phase LC/MS/MS system for the simultaneous analysis of drugs and their polar metabolites. J Sep Sci 2010;33:873–9.

[5] Maurer HH, Pfleger K, Weber AA. Mass spectral and GC data of drugs, poisons, pesticides, pollutants and their metabolites, 4th ed. Weinheim: Wiley-VCH; 2011.

[6] Broecker S, Herre S, Wust B, Zweigenbaum J, Pragst F. Development and practical application of a library of CID accurate mass spectra of more than 2,500 toxic compounds for systematic toxicological analysis by LC-QTOF-MS with data-dependent acquisition. Anal Bioanal Chem 2011;400:101–17.

[7] Tyrkko E, Pelander A, Ojanpera I. Differentiation of structural isomers in a target drug database by LC/Q-TOFMS using fragmentation prediction. Drug Test Anal 2010;2:259–70.

[8] Wissenbach DK, Meyer MR, Remane D, Weber AA, Maurer HH. Development of the first metabolite-based LC-MSn urine drug screening procedure – exemplified for antidepressants. Anal Bioanal Chem 2011;400:79–88.

[9] Wissenbach DK, Meyer MR, Weber AA, Remane D, Ewald AH, Peters FT, Maurer HH. Towards a universal LC-MS screening procedure – can an LIT LC-MS^n screening approach and reference library be used on a quadrupole-LIT hybrid instrument? J Mass Spectrom 2012;47:66–71.

[10] Remane D, Meyer MR, Wissenbach DK, Maurer HH. Full validation and application of an ultra high performance liquid chromatographic-tandem mass spectrometric procedure for target screening and quantification of 34 antidepressants in human blood plasma as part of a comprehensive multi-analyte approach. Anal Bioanal Chem 2011;400:2093–107.

2.25 Analytical quality in the Latin America area

Gabriel Alejandro Migliarino

Summary

We will address different aspects related to analytical quality in Latin America. When we talk about analytical quality, we refer to different links, which are related to each other and have the same goal: *to generate clinically useful results in patient healthcare.* Laboratories consider that analytical quality is assured and is the responsibility of manufacturers' assays. This is just a perception and there is no objective evidence supporting it. Laboratories are responsible for the analytical quality of their results and if they want to generate clinically useful results they must assure the analytical phase as well as the pre- and post-analytical phases [1]. From my point of view, there are three main issues that define the status of our countries: education, resources and guidances.

Within a region, we find laboratories of different sizes and complexity. In general, we can find a small number of high-complexity laboratories in the capitals and/or major cities. These laboratories operate under international standards [ISO, College of American Pathologists (CAP), etc.] with a strong emphasis on analytical quality. However, the majority of laboratories in the region experience very different situations. It is very common to find small laboratories and in some cases individual ones that, for different reasons, have deficiencies at an analytical level.

Regional regulations have different problems:

- inadequate and/or obsolete regulations;
- there is no effective enforcement of existing regulations; and
- there are regulations where analytical quality is not taken into account.

When a sample arrives at a laboratory or blood bank, it will undergo different phases: pre-analytical, analytical and post-analytical. These three phases will have to be supported by a quality management system with the purpose of obtaining clinically useful results. The laboratory or blood bank is responsible for assuring the three phases. If any of these phases fails, regardless of which one it is, the result will be the same. The results obtained will not be clinically useful.

Let us now focus on the analytical phase. Everything begins with an instrument qualification, where we make sure that the instruments have been properly installed and can operate as established by the manufacturer. Next, quality requirements must be established in order to know the amount of error we can allow in the assays, so that these errors do not nullify the clinical usefulness of the results obtained. Next, we have the method evaluation. This evaluation can have a validation or verification scope as the case may be, but in both cases it will entail the determination of total error in the assays by using statistical tools (protocols). Then, we will compare the total error obtained in the method evaluation with the previously established quality requirements to decide whether this method will be able to be used in the clinical laboratory or blood bank. Once we have accepted the method, we must make sure that this method we know to be good and acceptable in terms of time will continue to be so on a daily basis. This is where the internal and external statistical control comes in.

In many countries within the region, there is no direct representation of the firms that manufacture the instruments, and they work with distributors. There are no clearly defined guidelines for instrument qualification. Perhaps this is a problem that goes beyond the region. The seller and user responsibilities are not clear regarding this issue. Installation conditions and requirements are not previously agreed on. Qualifications are carried out many times, but due to the shared responsibility between users and sellers no corresponding records are generated. Once the instrument has been installed, deficiencies in instrument maintenance can be seen on the part of the user and sometimes on the part of the manufacturer. This is accentuated when there are inexperienced distributors involved. At a maintenance level, there are also deficiencies in terms of records and compliance.

In terms of health, it has always been very difficult to define and measure quality; therefore, the focus has been on performance. Quality and performance are related concepts but are not the same. Performance is a measure of how well we are doing something; quality is a measure of performance against a standard or requirement. Quality in terms of health has to do with the level of performance needed for proper medical care. In order to assess quality, and therefore control it, it is necessary to work with quality requirements. "Tolerance limits" (quality requirements) need to be defined in order to determine optimum or poor quality and therefore identify deficiencies. In an ideal world, it would be excellent to be able to refer to a single source based on what clinical usefulness is, where we could have access to a quality requirement for each of the assays. At the same time, we would also expect these requirements to be expressed in a single unit, for example in terms of how simple they are top handle, percentage accuracy, etc. There is great confusion in terms of which quality requirements source to use and how to deal with the issue of units. This is why, unfortunately, when in doubt many laboratories choose not to work with quality requirements due to a lack knowledge or information.

Method evaluation

When we think about method evaluation [2] the deficiencies are huge. Such evaluation is not a common practice. Laboratories do not know:

- what method evaluation is;
- why they should evaluate their methods;
- whether they should validate or verify their methods, when they should do it;
- which parameters they should validate/verify; or
- how they should validate or verify these parameters.

As you can see, uncertainty regarding this issue is considerable. There is also no clear information available in Spanish. In general, laboratories have no quality requirements – they think this activity will increase their costs and that they need to be experts in statistics in order to carry it out. It is clear that there is a misinformation regarding this issue. Therefore, only "special" laboratories in the region routinely implement method evaluation. Many of them already have, or are in the process of obtaining, the ISO 15189:2007 or CAP accreditation.

Internal and external quality control

We will now talk about internal and external quality control [3]. Many laboratories work with internal statistical quality control. However, there are many deficiencies in this matter. The same problems caused by identical causes can be seen in different countries within the region. Many laboratories do not handle control material properly, and there are inconsistencies in terms of storage temperature, reconstitution (volumetric instruments, diluents, and pipettes), freezing (containers), defrosting procedure and homogenization before processing.

Since working with quality requirements is not a common practice, many laboratories do not take into account the quality requirements when planning their internal statistical quality control scheme. If we consider more technical aspects, we will see there are serious problems with the mean and standard deviation assignment mechanisms in control graphs. Arbitrary means and standard deviation, which do not represent the method performance in the laboratory, are usually assigned to quality control graphs. Since quality control is not planned, taking into account the assay performance and quality requirements is completely arbitrary, as is the control rules selection, the amount of controls and the number analytical runs. We can also often see an incorrect statistical handling of the data, and serious problems with the records.

Many laboratories perform an internal statistical quality control because they are supposed to, but they do not use the information they obtain from it for the continuous improvement of their measurement procedures. This means that they spend money and time but they do not take advantage of the investment.

There are several peer group comparison schemes in the region. I believe these programs are essential to follow up on the analytical performance when estimating the bias and coefficient of variation on a monthly basis for comparison with the quality requirement and be able to assess the quality of our assays by estimating the statistical sigma level. This would be an indicator of analytical quality.

Taking into account the total number of laboratories in the region, we will see that only a few perform this type of program. Basically this happens due to a lack of resources, lack of knowledge or lack of interest. Likewise, it is important to emphasize that organizers of this type of scheme normally use a control material lot for the US and Canada, another lot for Europe, a different one for Asia and Oceania and another, the poorest one, for Africa and Latin America. This makes peer comparison groups small and therefore, the information obtained from these schemes weak.

Many laboratories perform external quality assessment schemes. Some of these schemes are national; others are regional; and others worldwide. It is not common to have regulatory external quality assessment schemes in countries of the region. The most common problems are:

- that many of the programs offer true values obtained by consensus;
- ignorance in many laboratories in relation to the traceability of their assays and therefore incorrect grouping in the schemes

- laboratories usually consider the last report and do not perform a retrospective evaluation of the data in search of trends or deviations
- laboratories consider the supplier's result classification without taking into account the quality requirements;
- due to a lack of knowledge and clear guidelines, they incorrectly use the external quality assessment schemes when estimating the bias by using a single survey;
- due to a lack of knowledge and/or information, reports are incorrectly interpreted
- there are poor comparison groups;
- late report delivery;
- problems with material delivery (import) and preservation; and,
- programs do not offer educational information or they do it poorly.

From my point of view, there are three main issues: education, recourses and guidance.

Talking about education, I have a great concern. Universities do not update their academic degree programs and they include few quality topics and even fewer analytical quality ones. Graduate programs available within the region usually involve quality management. They include few analytical quality topics and they are not health-oriented.

You can find courses on quality that include analytical quality but courses on quality control are more common. With the emergence of the ISO 15189:2007 there are now courses on traceability and uncertainty (although they are few). I must mention here that the concept of total error is deeply rooted in the region. With the advent of ISO 15189:2007, the participating laboratories (and only them), have started to estimate the uncertainty of their measurement processes. However, they only estimate it and do not use the information due to the lack of clear guidelines and experience.

Those of us who try to incorporate concepts (that are new to the region) are often considered to be talking about science fiction or, even worse, trying to force laboratories to do things they cannot do or are considered to be putting weird ideas into people's heads.

I am really not talking about anything new when I tell you that the economic situation in the region is complicated and in some countries even critical. Many professionals in the field of clinical laboratories or blood banks are trained to put the knowledge obtained in their jobs into analytical quality improvement. When they try to implement these improvements, they often clash with laboratory management regarding resources. This aversion is sometimes the result of ignorance due to the lack of understanding of the matter by the management and other times it means that it is impossible to make use of resources. I do not want this to be an excuse, but the health system in the region has not incorporated the quality concept and even less the analytical quality concept as requirements for the results; therefore they are not willing to pay for quality.

Conclusion

In the various countries within the region, there are different laboratories. These laboratories are usually accredited with or are in the process of obtaining the ISO 15189:2007, CAP or local or regional regulation accreditation.

To say that the analytical quality of our assays is ensured in the region is a mere perception. We need to prove this with objective evidence. There is still a long way to go but the situation is improving. The financial situation in Latin America is usually against us. Sometimes it is difficult to negotiate with management to obtain the resources to do what is needed. At other times, it is very difficult for management to administer resources as they do not have any.

References

[1] Westgard JO. Assuring the right quality right: good laboratory practices for verifying the attainment of the intended quality of test results. Madison Wisconsin: Westgard Inc.; 2007.
[2] Westgard JO. Basic method validation. 3rd ed. Madison Wisconsin: Westgard Inc.; 2008.
[3] Westgard JO. Basic QC practices. 3rd ed. Madison Wisconsin: Westgard Inc.; 2010.

2.26 Standardization in molecular diagnostics: definitions and uses of nucleic acid reference materials

Deborah A. Payne, François Rousseau, Cyril D.S. Mamotte, David Gancberg, Ron H.N. van Schaik, Heinz Schimmel and Mario Pazzagli

Summary

Molecular diagnostics is one of the most rapidly growing areas of laboratory medicine. This rapid growth of clinical molecular tests has outpaced the availability and development of reference methods and reference materials. The European Union, the United States Food and Drug Administration, National Institutes of Health, Organization for Economic Co-operation and Development, World Health Organization (WHO), Clinical Laboratory Standards Institute (CLSI) and numerous professional organizations (e.g., International Federation of Clinical Chemistry (IFCC) and the Association for Molecular Pathology) are examples of organizations that recognize a need for reference materials. Yet there is a lack of harmonization between the numerous international organizations [i.e., CLSI, WHO, International Vocabulary of Metrology (VIM) and the International Organization for Standardization (ISO)] that are currently either certifying or defining reference materials. Such methods and materials are important for the development, validation, instrument/method evaluation, assay set up, training and interpretation of diagnostic methods and tests. The objective of this paper is to review and clarify the definitions and applications for the use of reference materials in the context of molecular diagnostics.

An overview of definitions and applications for the use of reference materials

Molecular diagnostics is one of the most rapidly growing areas of laboratory medicine. This rapid growth of clinical molecular tests has outpaced the availability and development of reference methods and reference materials. The European Union, US Food and Drug Administration, US National Institutes of Health, the Organization for Economic Co-operation and Development, WHO, CLSI and numerous professional organizations [i.e., IFCC and Association for Molecular Pathology] are examples of organizations that recognize a need for reference materials. International organizations – i.e., CLSI, WHO, VIM and the ISO – differ in their definitions of various types of reference materials. According to the ISO, reference materials should be documented to be homogeneous and stable. Once these characteristics have been established, progressively more rigorous requirements determine which materials are considered quality control materials, calibrators and finally certified reference materials.

As mentioned previously, terminology differs between various organizations. This difference is illustrated by the inclusion of less rigorously tested reference materials listed by various websites and databases (▶Tab. 2.26.1). The Centers for Disease Control website in ▶Tab. 2.26.1 includes so called "lower level" or "lower order" reference materials that do not comply with the ISO requirement for documented homogeneity or stability. The IFCC Committee for Molecular Diagnostics supports the use of the ISO definitions for reference materials [1] but recognizes that "lower level materials" may be used by laboratories in the absence of appropriate certified reference materials.

Such materials are important for the development, validation, instrument/method evaluation, assay setup, training and interpretation of diagnostic methods and tests. For assay development, nucleic acid reference materials (NARMs) can be used to adjust technical parameters, such as extraction protocols, and the impact of various matrices on nucleic acid quality and assay results. For instance, a certified quantity of nucleic acid or viral particles spiked into matrices containing fecal material can be used to evaluate various extraction processes [2]. The presence of inhibitors are detected by quantitative polymerase chain reaction when diluted and undiluted samples yield similar quantitative values, while pure specimens would produce quantitative different results for diluted and undiluted samples.

NARMs can also be used to confirm instrument or methodology comparability. Before the advent of the WHO hepatitis C virus (HCV) International unit (IU), various platforms had different values for a given amount of HCV. For example, 1 IU of HCV corresponded to 0.9 copies/mL using the Amplicor manual method but 5.2 copies/mL for the Versant HCV version 3.0 method. By implementing the HCV NARM and a conversion factor, viral load results are commutable between platforms [3].

▶Fig. 2.26.1 illustrates the results of a bone marrow engraftment analysis and how NARMs can be used for training. The question arising from this example is "Should the director accept calculations performed by the technologist?". The answer is "No" because the informative peak in the 95% admixture control is within the background of the assay. STR NARMs (SRM 2390, NIST) would help establish a level of confidence with calculating percentages and assist with the training of technologists on acceptable vs. unacceptable results.

Lastly, NARMs could be used to assist in interpretation. For instance, interpreting the FMR1 expansions associated with fragile X can be difficult without standards (▶Fig. 2.26.2). Inclusion of the NARM for FMR1 (SRM 2399, NIST) facilitates interpretation by providing a reproducible numerical cut-off for interpretation (see ▶Fig. 2.26.2, right, showing a numerical cut-off).

Tab. 2.26.1: Website for databases listing sources of nucleic acid materials.

Organization	Website link
EuroGentest	www.eurogentest.org
CDC Genetic Testing Reference Materials*	http://wwwn.cdc.gov/dls/genetics/rmmaterials

*Uses the term "Reference Materials" for materials that are not compliant with the ISO definition for reference materials.

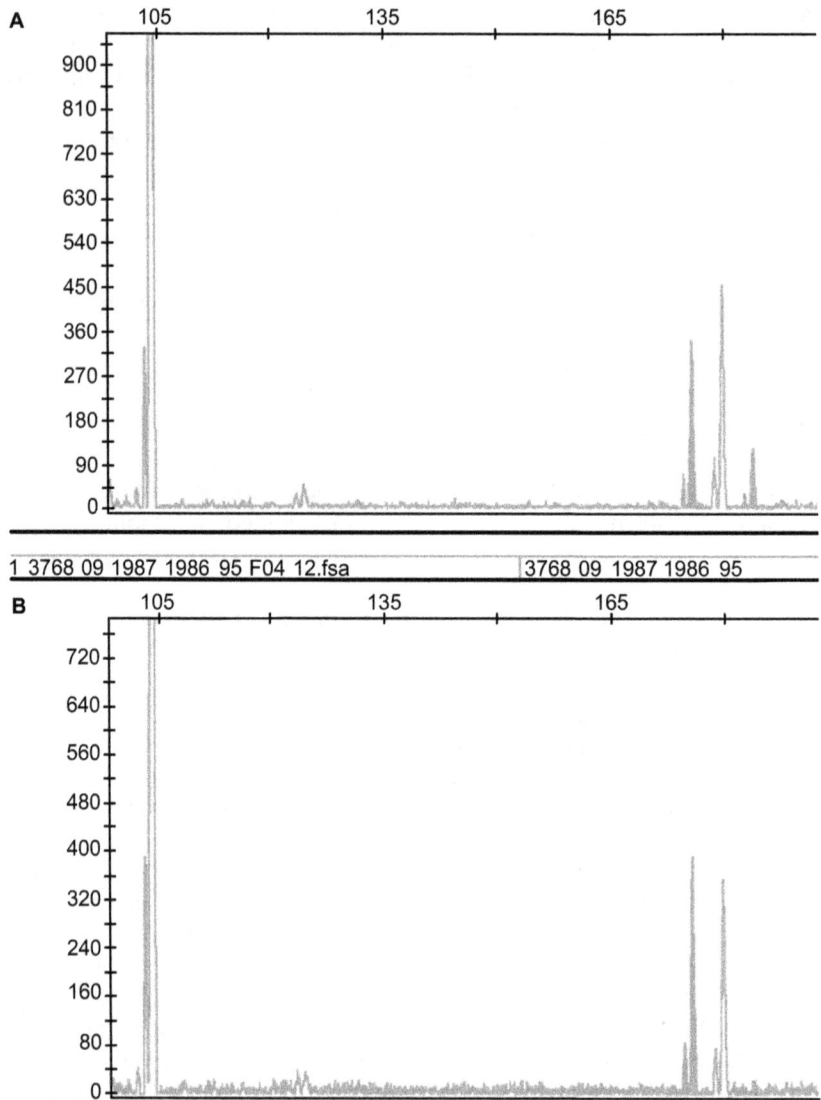

Fig. 2.26.1: A bone marrow chimera engraftment analysis using short tandem repeats.
(A) Raw patient data showing peak height for the informative allele [i.e. far right peak, peak height approximately 100 Relative Fluorescent Units (RFU) listed on the y-axis].
(B) A 95% admixture of donor and recipient. Note that the informative peak is below the background.

Conclusion

Numerous definitions of NARMs from various entities can cause confusion for end users. For example, the use the terms "lower level" or "lower order" reference materials are

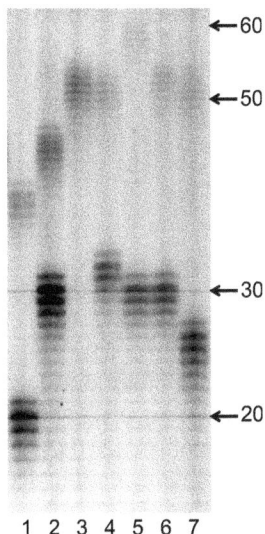

Fig. 2.26.2: Polymerase chain reaction and high-resolution acrylamide gel analysis of FMR1 alleles in the range between normal alleles and pre-mutations, which are more difficult to resolve on Southern blot.
The number of repeats in the expansion is shown at the right of the panel and was determined using a Nucleic Acid Reference Material (NARM).

misnomers in that these materials do not meet the basic requirement of reference materials as defined by the ISO. However, because of the relative scarcity of NARMs, lower level materials may need to be used to address various aspects of clinical molecular diagnostic testing. Specifically, NARMs are important for the development, validation, instrument/method evaluation, setup, training and interpretation of diagnostic methods and tests. For these reasons, NARMs can be used to improve the quality of laboratory testing, interpretation and ultimately patient care.

References

[1] Payne DA, Mamotte CD, Gancberg D, Pazzagli M, van Schaik RH, Schimmel H, Rousseau F, IFCC Committee for Molecular Diagnostics (C-MD). Nucleic acid reference materials (NARMs): definitions and issues. Clin Chem Lab Med 2010;48:1531–5.
[2] Petrich A, Mahony J, Chong S, Broukhanski G, Gharabaghi F, Johnson G, Louie L, Luinstra K, Willey B, et al, and for the Ontario Laboratory Working Group for the Rapid Diagnosis of Emerging Infection et al. Multicenter comparison of nucleic acid extraction methods for detection of SARS coronavirus RNA in Stool Specimens. J Clin Microbiol 2006;48:2681–8.
[3] Franciscus A, Highleymann L. Hepatitis C Support Project (HCSP) Fact Sheet, HCV Viral Load Tests, 2003, http://www.hcvadvocate.org/hepatitis/factsheets_pdf/viralload.pdf.

2.27 Profiling of antiphospolipid antibodies – association with cerebrovascular events in APS

Dirk Roggenbuck

Summary

Association of cerebrovascular events with antiphospholipid antibodies (aPL) in patients with antiphospholipid syndrome (APS) is reported controversially and depends on aPL reactivity and assay techniques. A novel line immunoassay (LIA) with phosphatidylserine, phosphatidylinositol, cardiolipin, and beta$_2$-glycoprotein I (β_2GPI) as phospholipid targets was used to detect aPL, IgG and IgM in 85 APS patients and 65 control individuals with APS-unrelated disease. Eight (9.4%) of the 85 APS patients had pregnancy morbidity, 62 (72.9%) had arterial and/or venous thrombosis, 57 (67.1%) had deep venous thrombosis and 18 (21.2%) systemic lupus erythematosus. Thirteen (15.3%) APS patients demonstrated cerebrovascular events comprising cerebral transient ischemic attack (TIA) and/or ischemic stroke. Anti-phosphatidylinositol IgM (3/10), anti-phosphatidylserine IgM (5/10) and anti-cardiolipin IgM (7/10) antibodies detected by the LIA demonstrated a significantly higher prevalence in the APS patients suffering from TIAs compared with the remaining APS patients (0/75, $p < 0.05$; respectively). The detection of 3 or more aPL IgM by LIA also revealed a significantly higher prevalence in this APS patient cohort (5/10 vs. 11/75; $p < 0.05$). Analyses of the 13 APS patients with cerebrovascular events revealed a significantly higher prevalence of anti-phosphatidylinositol IgM (3/13) and anti-cardiolipin IgM (9/13) compared with the remaining APS patients (0/72 and 24/72; $p < 0.05$; respectively). LIA technique is an interesting alternative to ELISA for aPL antibody detection and profiling in APS patients. Multiplex detection of aPL IgM by LIA revealed a significant association with cerebrovascular events in APS.

Introduction

State-of-the-art laboratory diagnosis of APS requires the running of several enzyme-linked immunosorbent assays (ELISAs) simultaneously in routine laboratories, creating a demand for multiplex tests detecting antiphospholipid (aPL) antibodies [1]. Line immunoassays or other multiplex techniques like biosensor analysis may provide the opportunity to detect several aPL antibodies simultaneously, as reported for multiplex assessment of autoantibodies [2,3]. Line immunoassays are an exciting alternative to ELISA for aPL antibody detection and profiling in APS patients. Antibody profiling can reveal further information regarding disease activity and outcome [4,5,6]. Thus, we investigated the clinical association in APS of aPL antibodies assessed by ELISA and a novel LIA.

Patients and methods

Eighty-five patients with APS (71 females and 14 males) were recruited for this study. Patients suffering from APS had been diagnosed by characteristic clinical and serological criteria [5]. Eight (9.4%) of the 85 patients with APS suffered from pregnancy morbidity, whereas 62 (72.9%) demonstrated arterial and/or venous thrombosis. Fifty-seven (67.1%) patients from the latter group suffered from deep vein thrombosis. Eighteen (21.2%) of the APS patients were additionally diagnosed as suffering from systemic lupus erythematosus. Thirteen (15.3%) APS patients showed transient ischemic attack (TIA) and/or ischemic stroke. Two APS patients, each with thrombotic events, met the diagnostic criteria for Sjögren's syndrome and scleroderma, respectively. One APS patient with thrombotic events suffered from rheumatoid arthritis, and 2 patients from spondyloarthropathies. As control groups, 65 patients (60 females and 5 males) suffering from diseases unrelated to APS (clinical symptoms suspicious of APS but without positive laboratory tests) and 79 sera from normal healthy, anonymous age- and sex matched donors, were included.

The study was approved by the local Ethical Committee (EA1/001/06). Written informed consent was obtained from each patient. All sera had been stored at –20°C.

Antibodies to cardiolipin (CL), beta$_2$-glycoprotein I (β_2GPI), phosphatidylserine (PS), and phosphatidylinositol (PI) in the patient sera were assessed by employing a commercially available LIA using purified human β_2GPI and phospholipids, according to the recommendations of the manufacturer (GA Generic Assays GmbH, Dahlewitz, Germany). The phospholipids CL, PS and PI and the protein β_2GPI were sprayed onto polyvinylidene difluoride (PVDF) membranes in lines for immobilization, as recently described for glycolipids [7]. For comparison, aPL antibodies to CL and β_2GPI were detected using a commercially available solid-phase ELISA (GA Generic Assays GmbH, Dahlewitz, Germany).

Results

Findings for the anti-β_2GPI and anti-CL IgG and IgM did not differ significantly according to the McNemar test ($p > 0.05$). Thus, the PVDF membrane employed in the novel LIA seems to induce the same conformational changes in the β_2GPI polypeptide.

Interestingly, the occurrence of aPL IgG antibodies to pure anionic phospholipids or to serum cofactors interacting with them detected by LIA appears to follow a particular pattern in APS (▶Tab. 2.27.1). The least frequently detected anti-PI IgG antibodies were always accompanied by anti-PS, anti-CL and anti-β_2GPI IgG. The more common anti-PS IgG occurred at significantly higher frequencies together with anti-CL and anti-β_2GPI IgG. These findings suggest the potential epitope spreading of aPL IgG from CL to PS, and further to PI with the involvement of β_2GPI reactivity. This phenomenon was not consistently found for aPL IgM. It remains to be shown whether this hypothetical epitope spreading observed for aPL IgG only affects anionic phospholipids. Further studies are warranted to investigate other phospholipid targets by LIA.

Regarding the clinical association in APS, only aPL antibodies detected by LIA and not by ELISA demonstrated a significantly higher prevalence in APS patients with

Tab. 2.27.1: Anti-phospholipid IgG profiles detected by line immunoassay investigating 85 patients with antiphospholipid syndrome.

n	anti-β_2GPI	anti-CL	anti-PS	anti-PI
2	+	–	–	–
9	–	+	–	–
1	–	+	+	–
2	+	+	–	–
20	+	+	+	–
7	+	+	+	+
44	–	–	–	–

β_2GPI, beta$_2$-glycoprotein I; CL, cardiolipin; PI, phosphatidylinositol; PS, phosphatidylserine.

Tab. 2.27.2: Clinical association with cerebrovascular events of antiphospholipid IgM antibodies detected by line immunoassay investigating 85 patients with antiphospholipid syndrome.

IgM	anti-CL	anti-β_2GPI	anti-PS	anti-PI	3 aPL IgM
TIA (n = 10)	p = 0.04		p = 0.001	p = 0.02	p = 0.02
TIA and/or stroke (n = 13)	p = 0.03		p = 0.003		p = 0.01

aPL, antiphospholipid; β_2GPI, beta$_2$-glycoprotein I; CL, cardiolipin; PI, phosphatidylinositol; PS, phosphatidylserine; TIA, transient ischemic attack.

clinical symptoms (▶Tab. 2.27.2). In particular, anti-PI IgM, anti-PS IgM and anti-CL IgM antibodies determined by the novel LIA showed a significant higher prevalence in the APS patients suffering from TIAs (n = 10) compared with the remaining APS patients (p = 0.001, p = 0.02 and p = 0.04, respectively). The assessment of 3 or more aPL IgM antibodies by LIA also demonstrated a higher prevalence in this APS patient cohort (p = 0.02). Analyses of the 13 APS patients suffering from ischemic stroke and/or TIA also revealed a higher prevalence of anti-PI IgM and anti-CL IgM antibodies (p = 0.003 and p = 0.03, respectively). In fact, the assessment of 3 or more aPL IgM antibodies by LIA again had a significantly higher prevalence in this APS patient group (p = 0.01).

Conclusion

According to the APS classification criteria, aPL antibody testing plays an essential role in diagnosing APS and leads to important treatment decisions [1,8]. Assessment of aPL still remains a diagnostic challenge [9,10]. Multiplex detection of aPL antibodies by LIA may address these issues and provide a reliable tool for aPL antibody profiling. Interestingly, only aPL IgM detected by LIA revealed a significant association with clinical symptoms in APS. In summary, LIA appears to be an alternative to ELISA for aPL antibody detection in the serological diagnosis of APS.

Competing interests

Dirk Roggenbuck has a management role at and is a shareholder of GA Generic Assays GmbH and Medipan GmbH. Both companies are diagnostic manufacturers.

References

[1] Miyakis S, Lockshin MD, Atsumi T, Branch DW, Brey RL, Cervera R, Derksen RH, DE Groot PG, Koike T, Meroni PL, Reber G, Shoenfeld Y, Tincani A, Vlachoyiannopoulos PG, Krilis SA. International consensus statement on an update of the classification criteria for definite antiphospholipid syndrome (APS). J Thromb Haemost 2006;4:295–306.
[2] Grossmann K, Roggenbuck D, Schroder C, Conrad K, Schierack P, Sack U. Multiplex assessment of non-organ-specific autoantibodies with a novel microbead-based immuno-assay. Cytometry A 2011;79:118–25.
[3] Metzger J, Von LP, Kehrel M, Buhl A, Lackner KJ, Luppa PB. Biosensor analysis of beta2-glycoprotein I-reactive autoantibodies: evidence for isotype-specific binding and differentiation of pathogenic from infection-induced antibodies. Clin Chem 2007;53:1137–43.
[4] Conrad K, Roggenbuck D, Reinhold D, Dorner T. Profiling of rheumatoid arthritis associated autoantibodies. Autoimmun Rev 2010;9:431–5.
[5] Pengo V, Biasiolo A, Pegoraro C, Cucchini U, Noventa F, Iliceto S. Antibody profiles for the diagnosis of antiphospholipid syndrome. Thromb Haemost 2005;93:1147–52.
[6] von Landenberg P, Scholmerich J, von Kempis J, Lackner KJ. The combination of different antiphospholipid antibody subgroups in the sera of patients with autoimmune diseases is a strong predictor for thrombosis. A retrospective study from a single center. Immunobiology 2003;207:65–71.
[7] Conrad K, Schneider H, Ziemssen T, Talaska T, Reinhold D, Humbel RL, Roggenbuck D. A new line immunoassay for the multiparametric detection of antiganglioside autoantibodies in patients with autoimmune peripheral neuropathies. Ann N Y Acad Sci 2007;1109:256–64.
[8] Cervera R, Khamashta MA, Shoenfeld Y, Camps MT, Jacobsen S, Kiss E, Zeher MM, Tincani A, Kontopoulou-Griva I, Galeazzi M, Bellisai F, Meroni PL, Derksen RH, de Groot PG, Gromnica-Ihle E, et al. Morbidity and mortality in the antiphospholipid syndrome during a 5-year period: a multicentre prospective study of 1000 patients. Ann Rheum Dis 2009;68:1428–32.
[9] Horstman LL, Jy W, Bidot CJ, Ahn YS, Kelley RE, Zivadinov R, Maghzi AH, Etemadifar M, Mousavi SA, Minagar A. Antiphospholipid antibodies: paradigm in transition. J Neuroinflammation 2009;6:3.
[10] Roggenbuck D, Egerer K, von Landenberg P, Hiemann R, Feist E, Burmester GR, Dorner T. Antiphospholipid antibody profiling – Time for a new technical approach? Autoimmun Rev 2012 doi:10.1016/j.autorev2012.02.01, in press.

2.28 Plasma levels of soluble CD30 and CD40L in pediatric patients after liver transplantation

Olga P. Shevchenko, Olga M. Tsirulnikova, Olga E. Gichkun, Rivada M. Kurabekova, Aleksandr A. Ammosov and Sergey V. Gautier

Summary

It has been found that the plasma levels of immunological markers such as soluble CD30 (sCD30) and soluble CD40L (sCD40L) are associated with graft dysfunction of the kidney, lung and heart. There are few reports about the relationship between these markers and some liver diseases [1,2,3]. The aim of the study was to find an association between plasma levels of sCD30 and sCD40L and graft dysfunction after living-donor liver transplantation (LDLT) in children with end-stage liver disease (ESLD). In children with ESLD pre-transplant plasma levels of sCD30 (78.3 ± 36.3 ng/mL) and sCD40L (3.3 ± 1.8 ng/mL) were significantly higher than in healthy donors (31.1 ± 11.7 ng/mL, $p < 0.01$ and 0.9 ± 0.6 ng/mL, $p < 0.01$, respectively). After LDLT, sCD30 and sCD40L plasma levels decreased significantly (56.4 ± 19.0 ng/mL, $p < 0.01$ and 2.7 ± 1.6 ng/mL, $p < 0.05$, respectively). The pre-transplant plasma level of sCD30 was 83.3 ± 34.1 ng/mL in children, who had graft dysfunction on days 26–32 after LDLT (n = 16). It was increased to 106.5 ± 15.9 ng/mL ($p < 0.05$) on days 21–30 after transplantation. Elevation of the sCD30 plasma level was observed 2–5 days prior to the increase in liver enzyme activity. In these children, pre-transplant plasma level of sCD40L was significantly higher (5.6 ± 2.0 ng/mL, $p < 0.05$) than in the other 49 patients. We concluded that elevated plasma levels of sCD40L before LDLT and sCD30 after LDLT were associated with graft dysfunction development in children after LDLT.

Introduction

Living-donor liver transplantation (LDLT) is a safe and lifesaving procedure that is now accepted as an established treatment in patients with end-stage liver disease. Soluble CD40L (sCD40L) is released from both stimulated lymphocytes and activated platelets and has been suggested as a link mediator between inflammation and thrombosis [4]. CD40L plays a pathogenic role in inflammatory disorders, such as autoimmune disease and allograft rejection [4,5,6].

CD30 is a membrane glycoprotein that belongs to the tumor necrosis factor superfamily. It is expressed on activated T cells. The monitoring of soluble CD30 (sCD30) levels may be useful for early diagnosis of an acute rejection episode [7,8,9].

The aim of the study was to evaluate the association between plasma levels of sCD30 and sCD40L and graft dysfunction after LDLT in children with end-stage liver disease.

Patients and methods

The study protocol was approved by the Ethics Committees at the Federal V. Shumakov Research Center of Transplantology and Artificial Organs. The study included 65 children with end-stage liver disease aged 14 ± 6 months and 38 adult living donors aged 37 ± 19 years. All recipients received 2- or 3-drug immunosuppressive therapy including tacrolimus or cyclosporine A. Blood samples for sCD30 and sCD40L measurements were collected before LDLT and at 21–30 days after LDLT. Plasma concentrations of sCD30 and sCD40L were measured by ELISA. The data were expressed as mean values plus or minus standard deviation. Correlations between variables were assessed using Spearman correlation analysis. Values of $p < 0.05$ were considered statistically significant. Calculations were performed with SPSS statistical pack V11.5.

Results

In children with end-stage liver disease, pre-transplant plasma levels of sCD30 (78.3 ± 36.3 ng/mL) and sCD40L (3.3 ± 1.8 ng/mL) were significantly higher than in healthy donors (31.1 ± 11.7 ng/mL, $p < 0.01$ and 0.9 ± 0.6 ng/mL, $p < 0.001$, respectively). Pre-transplant sCD40L and sCD30 plasma levels did not correlate with patient age, sex, total bilirubin, aspartate aminotransferase, alanine aminotransferase, gamma-glutamyl transferase, alkaline phosphatase, albumin, serum creatinine, plasma levels of C-reactive protein, carcinoembrionic antigen or alpha-fetoprotein ($p > 0.05$). No correlation was found among plasma sCD30, sCD40L and anti-human leukocyte antigens (HLA) class I, II antibodies.

After LDLT, sCD30 and sCD40L plasma levels decreased significantly (56.4 ± 19.0 ng/mL, $p < 0.01$ and 2.7 ± 1.6 ng/mL, $p < 0.05$, respectively). In the liver graft group, the patients who had graft dysfunction during the first month (n = 16) had a pre-transplant plasma level of sCD30 of 83.3 ± 34.1 ng/mL. It was increased to 106.5 ± 15.9 ng/mL ($p < 0.05$) on days 21–30 after transplantation. Elevation of the concentration of sCD30 was observed 2–5 days before an increase in liver enzyme activity (▶Fig. 2.28.1).

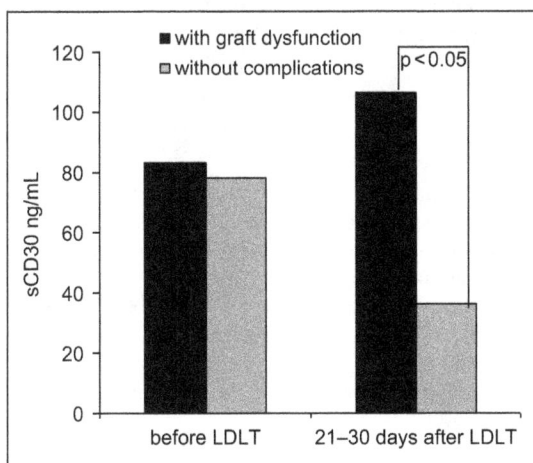

Fig. 2.28.1: Dynamic of sCD30 in patients with graft dysfunction and without complications before and after living-donor liver transplantation (LDLT).

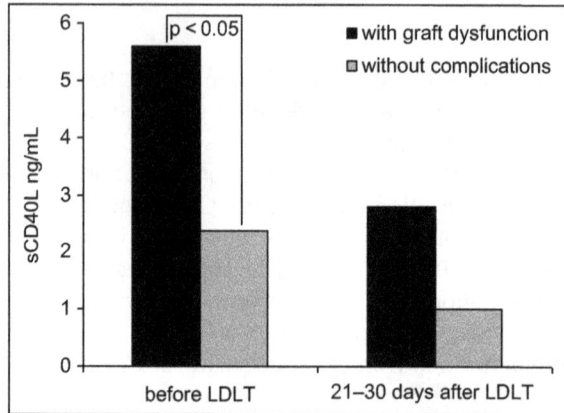

Fig. 2.28.2: Prognostic significance of sCD40L in patients with end-stage liver disease (ESLD) for the development of graft dysfunction after living-donor liver transplantation (LDLT).

In these children the pre-transplant plasma level of sCD40L was significantly higher (5.6 ± 2.0 ng/mL, $p < 0.05$) than in the other 49 patients (▶Fig. 2.28.2). In patients without complications, sCD30 and sCD40L plasma levels were 78.3 ± 26.3 ng/mL and 2.36 ± 1.14 ng/mL before transplantation and decreased significantly (36.5 ± 18.0 ng/mL, $p < 0.05$, and 1.00 ± 0.72 ng/mL, $p < 0.05$, respectively) after LDLT.

Conclusion

These data suggest that end-stage liver diseases are associated with increasing of sCD30 and sCD40L levels. In the 21–30 days after LDLT, plasma levels of sCD30 and sCD40L decreased significantly. A measurement of pre-transplant sCD40L concentrations might be useful to identify patients with end-stage liver disease at high risk for the development of graft dysfunction. Cases where the sCD30 level remained elevated in patients after LDLT might be prove to be an early predictor of graft dysfunction.

References

[1] Zheng YB, Gao ZL, Zhong F, Huang YS, Peng L, Lin BL, Chong YT. Predictive value of serum – soluble CD154 in fulminant hepatic failure. J Int Med Res 2008;36;728–33.
[2] Fábrega E, Unzueta MG, Cobo M, Casafont F, Amado JA, Romero FP. Value of soluble CD30 in liver transplantation. Transplant Proceed 2007;39:2295–6.
[3] Oertelt S, Invernizzi P, Selmi C, Podda M, Gershwin ME. Soluble CD40L in plasma of patients with primary biliary cirrhosis. Ann NY Acad Sci 2005;1051:205–10.
[4] Varo N, de Lemos JA, Libby P, Morrow DA, Murphy SA, Nuzzo R, Gibson CM, Cannon CP, Braunwald E, Schönbeck U. Soluble CD40L: risk prediction after acute coronary syndromes. Circulation 2003;108:1049–52.

[5] Kaya Z, Ozdemir K, Kayrak M, Gul EE, Altunbas G, Duman C, Kiyici A. Soluble CD40 ligand levels in acute pulmonary embolism: a prospective, randomized, controlled study. Heart Vessels, 14 April 2011, [epub ahead of print].

[6] Sternberg DI, Shimbo D, Kawut SM, Sarkar J, Hurlitz G, D'Ovidio F, Lederer DJ, Wilt JS, Arcasoy SM, Pinsky DJ, D'Armiento JM, Sonett JR. Platelet activation in the postoperative period after lung transplantation. J Thorac Cardiovasc Surg 2008;135:679–84.

[7] Süsal C, Döhler B, Sadeghi M, Salmela KT, Weimer R, Zeier M, Opelz G. Posttransplant sCD30 as a predictor of kidney graft outcome. Transplantation 2011;91:1364–9.

[8] Kennedy MK, Willis CR, Armitage RJ. Deciphering CD30 ligand biology and its role in humoral immunity. Immunology 2006;118:143–52.

[9] Truong DQ, Darwish AA, Gras J, Wieërs G, Cornet A, Robert A, Mourad M, Malaise J, de Ville de Goyet J, Reding R, Latinne D. Immunological monitoring after organ transplantation: potential role of soluble CD30 blood level measurement. Transpl Immunol 2007;17:283–7.

2.29 Diagnostic and prognostic value of presepsin (soluble CD14 subtype) in emergency patients with early sepsis using the new assay PATHFAST presepsin

Eberhard Spanuth, Henning Ebelt, Boris Ivandic and Karl Werdan

Summary

Background: Presepsin (soluble CD14 subtype, sCD14-ST) is a circulating molecule fragment derived from soluble CD14 (sCD14) and serves as mediator of lipopolysaccharide response against infectious agents. Initial evidence suggested that presepsin may be beneficial as a sepsis marker.

Methods: In 140 septic patients admitted to the emergency department and 119 healthy persons (control group), presepsin and procalcitonin levels were determined at admission and after 24 and 72 h. Primary endpoint was death within 30 days. The combined "major adverse event" (MAE) endpoint consisted of at least one of the primary or secondary endpoints (intensive care, mechanical ventilation or dialysis).

Results: Mean values of presepsin were 159 pg/mL (90% CI: 148–171) in the control group and 2,563 pg/mL (90% CI: 1458–3669) in the patient group. The 30-day mortalities were 3.5%, 25% and 67% in patients with low-grade sepsis, severe sepsis and septic shock, respectively. In contrast to procalcitonin, presepsin values differed highly significantly between sepsis grades ($p < 0.0001$) and were comparable to the clinical scores. The mortality increased from the 1st to the 4th quartile of presepsin, from 2.7% to 39.4%. Presepsin demonstrated superior prognostic accuracy for 30-day risk of death. The area under the receiver operating characteristics curve was 0.878 (95% CI: 0.801–0.934) compared to 0.668 (95% CI: 0.570–0.757) and 0.815 (95% CI: 0.709–0.895) for procalcitonin and Acute Physiology and Chronic Health Evaluation II scores, respectively. During the first 72 h patients with MAEs showed significantly increasing presepsin values, whereas the values in patients without MAEs were decreasing.

Conclusion: Presepsin demonstrated a strong relationship with disease severity and outcome. Presepsin values were related to the course of the disease. In contrast to PCT, presepsin provided more reliable prognosis and early prediction of 30-day mortality already at admission.

Introduction

CD14 is a glycoprotein expressed on the membrane surface of monocytes/macrophages and serves as a receptor for complexes of lipopolysaccharides and lipopolysaccharide binding protein, activating the toll-like receptor 4 specific pro-inflammatory signaling cascade against infectious agents. Simultaneously, CD14 is shed from the cell membrane forming soluble CD14 (sCD14), which can be found in serum with two different molecular weights [1]. Clinical studies have revealed elevated sCD14 levels in septic patients [2] and an association with mortality [3]. In summary, the results showed no incremental diagnostic information of sCD14 and a poor predictive value for patient outcome [4,5]. Recently another molecular fragment of sCD14 was discovered and named sCD14 subtype (sCD14-ST) or presepsin [6]. Plasma presepsin levels are associated with systemic inflammation triggered by bacterial infections. The initial results of clinical studies suggested that presepsin may be a promising diagnostic biomarker of systemic infections or sepsis [7,8,9].

Materials and methods

We report on an analysis of 140 patients admitted to the emergency department with clear signs or at least a strong suspicion of sepsis who participated in the ProFS study. Additionally, 119 healthy volunteers (60 females and 59 males, aged 21 to 69 years, with a mean age of 42 years) were included for determination of the reference interval of presepsin. Ethylendiamintetraacetic acid (EDTA) plasma samples were collected at admission, 24 h, and 72 h later.

Presepsin level was determined using the PATHFAST Presepsin assay (Mitsubishi Medience, Tokio, Japan). Procalcitonin level was measured using a luminescence immune assay (BRAHMS, Henningsdorf, Germany).

The primary endpoint of the study was death within 30 days. Secondary endpoints were transfer to an intensive care unit, a need for mechanical ventilation and a need for dialysis. The combined endpoint of a "major adverse event" was determined as an incidence of at least one of the primary or secondary endpoints.

Diagnostic efficacy

The presepsin values in 119 healthy volunteers revealed an upper reference limit of 320 pg/mL (90% CI: 238–335 pg/mL) without any influence of age and gender. Mean values were 159 pg/mL (90% CI: 148–171) compared with 2,563 pg/mL (90% CI: 1458–3669) in the septic patients (n = 40). Although procalcitonin exhibited a significant difference (p = 0.01) between patients with low grade sepsis (n = 85) and severe sepsis or septic shock (n = 55), the discrimination of presepsin was superior, reaching a significance level of p < 0.0001 that was comparable to those of the clinical scores. This finding was confirmed by receiver operator characteristics (ROC) analysis for presepsin and procalcitonin and revealed areas under the curve of 0.71 (95% CI: 0. 62–0.78) and 0.64 (95% CI: 0.55–0.72), respectively.

Prediction of risk of mortality

Presepsin but not procalcitonin was strongly correlated with mortality. Quartiles of presepsin showed a strong association (p < 0.0001 for trend analysis) with the risk of 30-day death; whereas quartiles of procalcitonin did not (▶Tab. 2.29.1).

Median values of presepsin at admission in survivors and non-survivors were 823 pg/mL (95% CI: 678–973) and 2,124 pg/mL (95% CI: 1206–3604), respectively. The difference was highly significant (p < 0.0001) and comparable to those of the clinical

Tab. 2.29.1: Association between quartiles of presepsin or PCT and mortality in septic patients.

Quartile	1st (n = 37)	2nd (n = 35)	3rd (n = 35)	4th (n = 33)
Presepsin, pg/mL	177–512	524–927	950–1,810	1,850–15,757
Mortality (p < 0.0001)	2.7%	8.6%	17.1%	39.4%
PCT, ng/mL	0.10–0.38	0.39–1.73	1.76–7.0	8.1–292
Mortality (p = 0.9816)	26.7%	8.1%	8.3%	24.3%

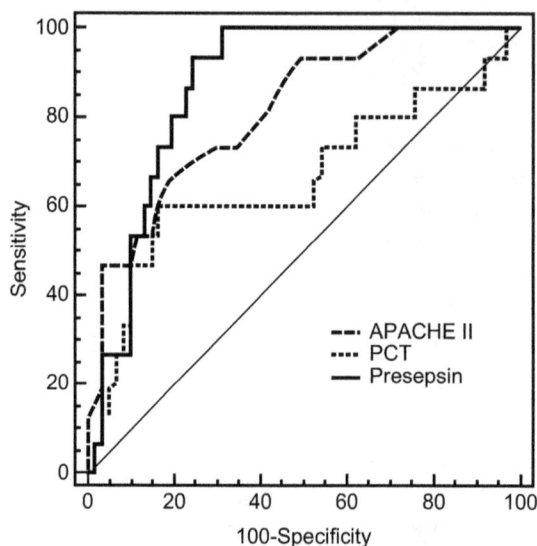

	Presepsin	PCT	APACHE
AUC, (95% CI)	0.878 (0.808-0.941)	0.668 (0.496-0.840)	0.815 (0.696-0.933)
p value	< 0.0001	= 0.0556	< 0.0001
Optimal cutoff points	1622 pg/mL	13.43 ng/mL	23
Sensitivity, % (95% CI)	93.3 (68.1–99.8)	60.0 (32.3–83.7)	66.7 (38.4–88.2)
Specificity, % (95% CI)	72.5 (62.2–81.4)	80.2 (70.6–87.8)	80.3 (68.2–89.4)
Positive Predictive Value, %	35.9	33.3	45.5
Negative Predictive Value, %	98.5	92.4	90.7

Fig. 2.29.1: Receiver operator characteristics curves of presepsin, PCT and APACHE scores for predicting 30-day mortality. The area under the curve for presepsin and APACHE scores were significantly >0.5. The optimal cutoff points along with their predictive values for the 30-day mortality of septic patients with PCT values ≥0.5 ng/mL are listed in the table.

scores. In contrast, the procalcitonin values showed no significant difference between survivors and non-survivors. The median values of procalcitonin were 1.84 ng/mL (95% CI: 1.23–2.73) and 2.07 ng/mL (95% CI: 0.31–21.23), respectively (p = 0.7452). ROC curve analyses were performed comparing the accuracy for the prediction of 30-day death of presepsin, procalcitonin and Acute Physiology and Chronic Health Evaluation II (APACHE II) scores in a subgroup of patients with procalcitonin values above the diagnostic threshold of 0.5 ng/mL for sepsis (n = 106). Presepsin demonstrated superior prognostic accuracy. The results are shown in ▶Fig. 2.29.1.

Tab. 2.29.2: Improvement of risk of 30-day mortality prediction by combining clinical scores and presepsin.

	AUC alone	AUC with presepsin	Net reclassification
APACHE II	0.815	0.905	21.05 + 33.33 = 54.38%
GCS	0.763	0.931	03.58 + 73.33 = 76.91%
MEDS	0.819	0.936	22.67 + 40.00 = 62.67%
SOFA	0.747	0.917	22.45 + 33.33 = 55.75%

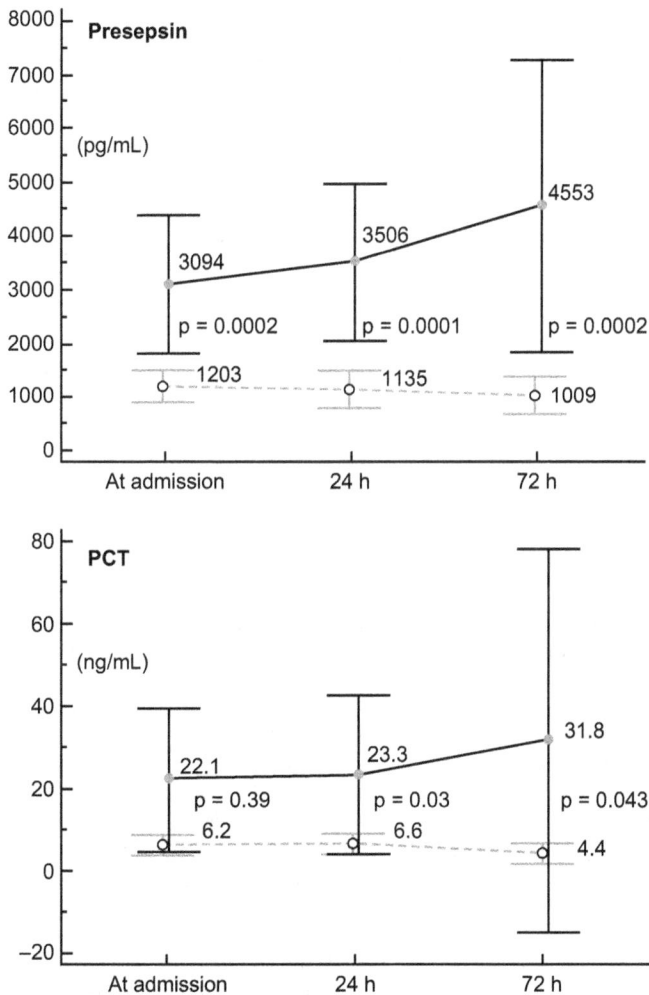

Fig. 2.29.2: Course of mean values of presepsin and PCT (error bars: 95% confidence interval) in patients who developed 30-day MAEs (solid line) and patients with favorable outcome without MAEs (dashed line).

The additive contribution of presepsin to a risk model based on the clinical scores was examined. The area under the curve of the APACHE II scores increased from 0.815 to 0.905 in the extended model. The net reclassification index was 21.05% + 33.33% = 54.38%, indicating that 21% more living patients are re-classified in the downward right direction and 33.3% more dead patients are re-classified in the upward right direction (▶Tab. 2.29.2). Similar results were obtained for the Glasgow Coma Score, the Mortality in Emergency Department Sepsis, and the Sequential Organ Failure Assessment scores, whereas procalcitonin failed to contribute significantly to the risk model (the p value of procalcitonin as an additional covariate was 0.2518).

Disease monitoring

All study patients received microbial therapy at admission when the clinical diagnosis of sepsis was established. To investigate the ability to determine the efficacy of initial therapeutic measures at a very early time point, presepsin and procalcitonin were measured at presentation, at 24 h and at 72 h after admission. In most patients with a favorable outcome without the occurrence of major adverse events within 30 days after admission (n = 104), the marker levels decreased between baseline and 72 h. In the patient group that experienced adverse outcomes (n = 36), both markers showed a tendency to increase in level. This effect was more pronounced for presepsin and becomes clear in the course of mean values, as shown in ▶Fig. 2.29.2.

Conclusion

Our findings indicate that presepsin may be a promising early diagnostic and prognostic marker in patients with sepsis. Presespin provided an early prediction of the risk of mortality and adverse outcomes at admission. The use of presepsin may improve the management of septic patients.

References

[1] Labeta MO, Durieux JJ, Fernandez N, Herrmann R, Ferrara P. Release from a human monocyte cell line of two different soluble forms of the lipopolysaccharide receptor, CD14. Eur J Immunol 1993;23:2144–51.

[2] Landmann R, Reber AM, Sansano S, Zimmerli W. Function of soluble CD14 in serum from patients with septic shock. J Infect Dis 1996;173:661–68.

[3] Burgmann H, Winkler S, Locker GJ, et al. Increased serum concentration of soluble CD14 is a prognostic marker in Gram-positive sepsis. Cin Immunol Immunopathol 1996;80:307–10.

[4] Herrmann W, Ecker D, Quast S, et al. Comparison of procalcitonin, sCD14 and interleukin-6 values in septic patients. Clin Chem Lab Med 2000;38:41–46.

[5] Pavcnik-Arnol M, Hojker S, Derganc M. Lipopolysaccharide-binding protein, lipopolysaccharide, and soluble CD14 in sepsis of critically ill neonates and children. Intensive Care Med 2007;33:1025–32.

[6] European Patent 1571160 assigned to Mochida Pharmaceutical Co., LTD. Assay kit and antibody for human low molecular weight CD14. Patent granted May 10, 2005.

[7] Shirakawa K, Furusako S, Endo S, et al. New diagnostic marker for sepsis: soluble CD14 subtype. Critical Care 2004;8(Suppl 1):P191doi: 10.1186/cc2658.

[8] Yaegashi Y, Shirakawa K, Sato N, et al. Evaluation of a newly identified soluble CD14 subtype as a marker for sepsis. J Infect Chemother 2005;11:234–38.

[9] Shozushima T, Takahashi G, Matsumoto N, et al. Usefulness of presepsin (sCD14-ST) measurements as a marker for the diagnosis and severity of sepsis that satisfied diagnostic criteria of systemic inflammatory response syndrome. J Infect Chemother 2011, published online 12 May.

[10] American College of Chest Physicians/Society of Critical Care Medicine Consensus Conference: definitions for sepsis and organ failure and guidelines for the use of innovative therapies in sepsis. Crit Care Med 1992;20:864–74.

[11] Ruiz-Alvarez MJ, Garcia-Valdecasa S. De Pablo R, et al. Diagnostic efficacy and prognostic value of serum procalcitonin concentrations in patients with suspected sepsis. J Intensive Care Med 2009;24:63–73.

[12] Pencina MJ, D'Agostino RB, D'Agostino BJ, Vasan RS. Evaluating the added predictive ability of a new biomarker: from area under the ROC curve to classification and beyond. Stat Med 2008;27:157–72.

[13] Tang BMP, Eslick GD, Craig JC, McLean AS. Accuracy of procalcitoin for sepsis diagnosis in critical ill patients: systemic review and meta-analysis. Lancet Infect Dis 2007;7:210–17.

2.30 Diagnostic workup of primary aldosteronism

Michael Stowasser

Summary

Because primary aldosteronism (PA) is common and associated with morbidity that is greater than with other forms of hypertension but abrogated by specific surgical and medical treatment, all hypertensive patients should be considered for testing for this condition. Optimal detection requires that (1) the aldosterone:renin ratio (ARR) is used to screen for PA; (2) potential confounders (e.g., medications, plasma potassium levels, dietary sodium, posture and time of day) are controlled or their effects considered when interpreting results; and (3) suppression testing is performed in those with repeatedly high ARR to definitively confirm or exclude PA. For the glucocorticoid-remediable form of PA, genetic testing has greatly facilitated detection. Adrenal venous sampling is the only reliable means of differentiating unilateral (surgically correctable) from bilateral (usually medically treated) PA. New, high-throughput mass spectrometric aldosterone assays have proven highly accurate and reproducible.

Importance of the detection of primary aldosteronism

Primary aldosteronism (PA) is much more common than previously thought, accounting for up to 5%–10% of hypertensive cases with most patients being normokalemic [1,2]. Aldosterone excess has adverse consequences independent of hypertension development, so that patients with PA have higher rates of cardiovascular and renal morbidity than those with other forms of hypertension [3]. As specific surgical (unilateral adrenalectomy) and medical (aldosterone antagonist) treatment effectively abrogates the morbidity associated with PA [4] and improves quality of life [5], this condition should be systematically sought and specifically treated.

Detection and diagnosis

The aldosterone:renin ratio (ARR) is the most reliable method of screening for PA but is not without false positives and negatives, which are often caused by medications (▶Tab. 2.30.1). Dietary salt restriction, concomitant malignant or renovascular hypertension, pregnancy and treatment with diuretics (including spironolactone), dihydropyridine calcium-channel blockers, angiotensin-converting enzyme inhibitors and angiotensin II receptor-antagonists can all lead to false-negative ratios by stimulation of renin secretion [6]. We recently found treatment with selective serotonin reuptake inhibitor antidepressants to lower the ARR [7]. Plasma potassium is a chronic regulator

Tab. 2.30.1: Examples of medications which can affect the aldosterone/renin ratio.

Medications that lower the ratio (potential for false negatives)	Medications that raise the ratio (potential for false positives)
Diuretics (including spironolactone)	Beta-blockers
Dihydropyridine calcium-channel blockers	Alpha-methyldopa
Angiotensin-converting enzyme inhibitors	Clonidine
Angiotensin-receptor blockers	Non-steroidal anti-inflammatory drugs
Selective serotonin reuptake inhibitors	Oral contraceptives containing ethinylestradiol and drosperinine*

*Only applies if renin is measured by direct active renin concentration, and not plasma renin activity.

of aldosterone secretion, so false negatives may also occur in the setting of uncorrected hypokalemia.

Beta- blockers, alpha-methyldopa, clonidine and non-steroidal anti-inflammatory drugs suppress renin and have the potential to cause false-positive ratios [6]. False positives may also be seen in patients with impaired renal function or advancing age. We have recently shown [8,9] that:

- females have higher ratios than males;
- false positives can occur during the luteal phase of the menstrual cycle or while taking an oral ethinylestradiol/drospirenone contraceptive preparation, but only if renin is measured as direct renin concentration and not plasma renin activity, and
- subdermal insertion of an implantable form of etonogestrel did not affect the ARR.

Where feasible, diuretics should be ceased for at least 6 weeks and other interfering medications for at least 2 weeks (preferably 4) before measuring the ratio, substituting other medications that have a lesser effect on results, such as verapamil slow-release, hydralazine and prazosin. Hypokalemia should be corrected and the patient encouraged to follow a liberal salt diet before ratio measurement.

Sensitivity is maximized by collecting blood midmorning from seated patients who have been upright (sitting, standing or walking) for 2–4 h. The ratio should be regarded as a screening test only. The test should repeated at least once (serially if conditions of sampling, including medications, are being altered) before deciding whether or not to go on to a reliable suppression test (e.g., fludrocortisone suppression testing) in order to definitively confirm or exclude PA [9,10,11].

Differentiation of subtypes

Computed tomography frequently misses aldosterone-producing adenomas and yet detects non-functioning nodules, so the only reliable means of differentiating unilateral (surgically correctable) from bilateral (usually treated medically) forms of PA is by adrenal venous sampling [10,11,12]. Restricting the procedure to 1 or 2 dedicated and experienced radiologists, and to a small number of referral centers (thereby ensuring a high throughput) helps to optimize the success rates of adrenal venous cannulation. For the glucocorticoid-remediable familial form of PA (familial hyperaldosteronism type I), genetic testing for the causative "hybrid" 11beta-hydroxylase/aldosterone synthase gene has greatly facilitated detection [13,14].

Assay considerations

It is important to recognize the potential inaccuracies of many currently available methods for aldosterone and renin (except in very well established and experienced laboratories). New, high-throughput mass spectrometric methods of measuring aldosterone have proven highly reliable and reproducible and represent a major step forward [15]. Validation of new assays of plasma renin activity using similar technology is awaited with interest.

Conclusions

Detection and specific treatment of PA not only leads to cure or at least the improvement of hypertension, but also reverses the excess morbidity and reduced quality of life associated with this condition. Careful attention to potentially confounding factors and use of high-quality assay methodologies improves the reliability of ARR results and assists in interpretation when screening for PA. Suppression testing permits a definitive confirmation of the diagnosis and, alternatively, exclusion of patients without PA from further, invasive testing. Adrenal venous sampling is the most reliable means of differentiating unilateral from bilateral forms and, although technically demanding, high-cannulation success rates can be achieved in centers with high throughput, provided the procedure is limited to a small number of dedicated radiologists.

References

[1] Gordon RD, Klemm SA, Tunny TJ, Stowasser M. Primary aldosteronism: hypertension with a genetic basis. Lancet 1992;340:159–61.
[2] Mulatero P, Stowasser M, Loh K-C, Fardella CE, Gordon RD, Mosso L, Gomez-Sanchez CE, Veglio F, Young Jr WF. Increased diagnosis of primary aldosteronism, including surgically correctable forms, in centers from five continents. J Clin Endocrinol Metab 2004;89:1045–50.
[3] Milliez P, Girerd X, Plouin PF, Blacher J, Safar ME, Mourad JJ. Evidence for an increased rate of cardiovascular events in patients with primary aldosteronism. J Am Coll Cardiol 2005;45:1243–8.
[4] Catena C, Colussi G, Nadalini E, Chiuch A, Baroselli S, Lapenna R, Sechi LA. Cardiovascular outcomes in patients with primary aldosteronism after treatment. Arch Intern Med 2008;168:80–5.
[5] Sukor N, Kogovsek C, Gordon RD, Robson D, Stowasser M. Improved quality of life, blood pressure, and biochemical status following unilateral laparoscopic adrenalectomy for unilateral primary aldosteronism. J Clin Endocrinol Metab 2010;95:1360–4.
[6] Stowasser M, Gordon RD. The aldosterone-renin ratio for screening for primary aldosteronism. The Endocrinologist 2004;14:267–76.

[7] Ahmed AH, Calvird M, Gordon RD, Taylor PJ, Ward G, Pimenta E, Young R, Stowasser M. Effects of two selective serotonin reuptake inhibitor antidepressants, sertraline and escitalopram, on aldosterone/renin ratio in normotensive depressed male patients. J Clin Endocrinol Metab 2011;96:1039–45.

[8] Ahmed AHA, Gordon RD, Taylor PJ, Ward G, Pimenta E, Stowasser M. Are women more at risk of false-positive primary aldosteronism screening and unnecessary suppression testing than men? J Clin Endocrinol Metab 2011;96:E340–6.

[9] Ahmed AH, Gordon RD, Taylor P, Ward G, Pimenta E, Stowasser M. Effect of contraceptives on aldosterone/renin ratio may vary according to the components of contraceptive, renin assay method and possibly route of administration. J Clin Endocrinol Metab 2011;96:1797–804.

[10] Stowasser M, Gordon RD, Rutherford JC, Nikwan NZ, Daunt N, Slater GJ. Diagnosis and management of primary aldosteronism. J Renin Angiotensin Aldosterone System 2001;2:156–69.

[11] Funder JW, Carey RM, Fardella C, Gomez-Sanchez CE, Mantero F, Stowasser M, Young Jr WF, Montori VM. Case detection, diagnosis, and treatment of patients with primary aldosteronism: an Endocrine Society clinical practice guideline. J Clin Endocrinol Metab 2008;93: 3266–81.

[12] Young WF, Stanson AW, Thompson GB, Grant CS, Farley DR, van Heerden JA. Role for adrenal venous sampling in primary aldosteronism. Surgery 2004;136:1227–35.

[13] Lifton RP, Dluhy RG, Powers M, Rich GM, Cook S, Ulick S, Lalouel JM. A chimaeric 11beta-hydroxylase/aldosterone synthase gene causes glucocorticoid-remediable aldosteronism and human hypertension. Nature 1992;355:262–5.

[14] Jonsson JR, Klemm SA, Tunny TJ, Stowasser M, Gordon RD. A new genetic test for Familial Hyperaldosteronism Type I aids in the detection of curable hypertension. Biochem Biophys Res Comm 1995;207:565–71.

[15] Taylor PJ, Cooper DP, Gordon RD, Stowasser M. Measurement of aldosterone in human plasma by semiautomated HPLC-tandem mass spectrometry. Clin Chem 2009;55:1155–62.

2.31 What should the clinical laboratory and the toxicologist-pharmacologist offer the poisoned patient?

Donald R.A. Uges

Summary

In the past, clinical toxicology has primarily been related to suicide attempts. Nowadays, drug and toxic substances are less available and people are increasingly inclined to attempt suicide mechanically (train, shooting, hanging) instead of chemically. Unknown recreational drugs or unknown substances from the Internet are taken for excitement and self-care, or given to victims for drug-facilitated crimes. Some individuals use poisons to attract attention, like Munchausen syndrome (by proxy). An increasing number of severe overdoses are caused by iatrogenic poisoning. Patients are becoming more vulnerable due to age (neonates, seniors), transplantation, severe illness, a combination of diseases and pharmacotherapies, congenital or acquired organ failure, or different CYP 450 status. It is obvious that the toxicologist-pharmacologist has to recognize the different possible causes of the clinical picture of the patient. Knowing only the conventional substances used in suicide attempts and classic drugs of abuse is not sufficient. It is clear that therapeutic drug monitoring and clinical toxicology are merging. It is important to choose the right qualitative and quantitative analytical methods [1]. Modern liquid chromatography with triple quad mass selective detection is offering the clinical laboratory, and the patient indirectly, a fantastic tool to exclude or include a therapeutic or toxic substance and to clarify the poisoning within a very short time. However, analytical tools are not always sufficient. In this case, creative and experienced clinical toxicologists, pharmacologists, clinicians and analytical chemists are required to solve most strange cases.

Introduction

If a patient (man/woman) is admitted to the hospital the first question will be: why did he (*read: he or she*) come to the emergency department?

Information from the patient (is possible and reliable), his general practitioner, family, police, place he was found (industry, on the street, in bed) is crucial when considering or excluding a possible poisoning. The patient might suffer from a real poisoning, a have subtherapeutic drug level or an apparent poisoning. In cases of anticonvulsants (e.g., phenytoin), both a too low or too high a blood level can cause a status epilepticus.

There are many apparent poisonings, such as hypoglycaemic coma; cerebrovascular accident, exhaustion (after prolonged severe seizures by strychnin), brain damage/brain death, meningitis, flash back or withdrawal symptoms (after drugs of abuse), (non) deliberate simulation (Munchausen by proxy or psychiatric disorder), idiosyncrasy/allergic reactions, unexpected symptoms of a disease (Lyme, Pfeiffer), drug levels in the blood that are too low (e.g., phenytoin or an antipsychotic). The clinician, pharmacologist and the lab can exclude an overdose in these instances.

The emergency physician will consult a clinical toxicologist about the information he got and the toxidromes. These clinical symptoms are essential for the successful recognition of poisoning patterns. A toxidrome is the collection of signs and symptoms that suggest a specific class of poisoning. The most important toxidromes are sympathomimetic, anticholinergic, hallucinogenic, opioid, sedative/hypnotic, cholinergic, withdrawal and serotonergic.

> Case: Patient with anticholinergic toxidrome: male, 27 years old with somnolence, slurred speech, combative behaviour. His sister said "He showed me a handful of pills, he intended to take". His medication is unknown. No other pills were found. His vital signs were: temp 37°C; heart rate 120 beats per minute, blood pressure 100/60 mmHg; respiration rate 22 breaths per minute. Skin warm; dry armpit. Pupils dilated, reactive.

Did he really take the tablets and which ones?

> First possibility (dry armpit): tricyclic antidepressants. The lab starts a therapeutic drug monitoring screen on amitriptyline + imipramine + clomipramine. In his serum 900 µg/L amitriptyline and 200 µg/L nortriptyline was found. The patient had collected his amitriptyline pills over several weeks, without therapeutic drug monitoring.

A partial toxidrome is pinpoints (extreme narrow pupils), which is well-known for opiates but organophosphates (parathion) and quetiapine (atypical antipsychotic) have the same partial toxidrome.

At an increasing number of nurses in the emergency department perform quick tests on exogenous substances. It started with a breath test for alcohol; however, the problem is that many patients are not capable of breathing into a breathalyzer and so the toxic methanol and (di-)ethylene glycol will be missed. In our hospital we started with a saliva test for drugs that are abused and benzodiazepines. The test is sent to our laboratory with an extra saliva sample for confirmation the next day and as quality control. However, the physician always sends edetate and clotted blood to our laboratory for general toxicological screening. We seldom get urine as our clinician does not routinely catheterize patients. Urine provides information from the patient before and not while at the emergency department. Maybe in the near future oral fluid will be the matrix of choice.

For the clinical laboratory and the toxicologist it is extremely important to get sufficient and useful information about the patient. A routine total toxicological screen is possible, but it is time consuming, expensive and even then many substances are missed. By using information as an analytical tool, we could bring speed and useful toxicological results closer together. With immunoassays on serum we can analyse paracetamol, lithium, ethanol, salicylates and tricyclic antidepressants.

We are using GC (head space) for the qualitative and quantitative determination of substances such as ethanol, methanol, acetone, ethylene glycol and gamma hydroxybutyric acid (GHB). Keep in mind that several doctors cannot reliably smell alcohol.

In the Netherlands the hospital pharmacy laboratories have a combination of high-performance liquid chromatography with diode array detection (HPLC-DAD) and library at their disposal, which is useful for the semi-quantitative screening on over 600 toxic drugs. As many clinical poisoning are iatrogenic, we have a large therapeutic drug monitoring service (www.bioanalysis.umcg.nl) [2].

If we have sufficient information about the patient from the emergency physician, we prefer to perform more purposeful determinations on one of our routine pieces of liquid chromatography-mass spectrometry/mass spectrometry (LC-MS/MS or triple quad) apparatus. For instance, if we are thinking about the possibility of an overdose with an antipsychotic, we measure all the available antipsychotics and their metabolites. For this purpose we have identical LC-MS/MS machines with a program that automatically chooses one of the eight columns, optimal gradient eluents, the collision energy and the drug and transition masses. For all the drugs we use the same serum volume (100 µL), internal standard (cyanoimipramine) and extraction method (protein precipitation). For quantitative analysis we use self-prepared lyophilized serum samples and quality control standards. If needed, we can perform a metal screen on induced coupled plasma technique (ICP/MS).

In special cases, such as Munchausen by proxy, drug-facilitated crime, strange illicit substances, drugs from the Internet, agricultural poisonings or when we really do not know which poison might be involved, we use the gas chromatography with mass selective detection (GC/MS) with extraction after derivatization, using a comprehensive library.

Normally, the laboratory must have a peer-reviewed quality system with certification or accreditation. The laboratory has to validate all of the assays. However, in clinical toxicology the patient goes first. So, if no validated determination is available, the head of the laboratory has to decide what is better to give a non-validated result or no result. The toxicologist and the physician have to discuss this problem. Recently we got green urine from a man with 33% methemoglobin. The green colour was caused by the antidote methylene blue that was administered. We thought about poppers (amyl nitrite) bought by the patient on Internet, and quickly (that evening) set up an assay by GC/MS, with positive results.

Today, we are increasingly seeing iatrogenic poisoning. Mistakes can be made by every (para-)medical professional. Therefore the medical laboratory needs to be able to measure more than just human matrices, such as blood, urine, saliva or liquor.

Case: A patient should only get a morphine infusion after his surgery. At once his glucose level decreased dramatically. Therefore the physician sent us the medication to exclude erroneous administration of insulin instead of morphine. The infusion conformed with the label. Later on, we learnt his blood for glucose testing was taken from his necrotic finger instead from his arm vein.

Case: In vacuoles in the swollen underside of a leg of a transgender patient we found silicone oil by ICP/MS. This oil came from his hip, which had been augmented to obtain a more female form.

All emergency physicians, nurses, toxicologists and technicians with several years working in the field of clinical toxicology have been confronted with bizarre cases that require experience, creativity, up-to-date knowledge and sometimes just luck.

Conclusion

The emergency physician, nurse, toxicologist, hospital pharmacist and chemist have to work closely together. If a severely ill patient arrives in the hospital, he requires optimal treatment as quickly as possible. Some antidotes and treatments can be life-saving for one patient, but might be disastrous for another. Every poisoning might be unique and requires the view of different medical professionals. The capability of the medical laboratory has to be in proportion to the needs, quality and kind of patients in that particular hospital.

Therapeutic drug monitoring, clinical toxicology and pharmaceutical analyses are merging. A combination is often needed. The introduction of the LC-MS/MS has dramatically improved the analytical possibilities of modern hospital laboratories. However, the toxicologist has to continuously study to keep up with the newest drugs, trends and possibilities with regards to poisoning. The physician has to know what the laboratory can offer him and to trust the sometimes unexpected results.

References

[1] Uges DRA. Medical toxicology. In: Moffat AC, Osselton D, Widdop M. Brian Clarke's analysis of drugs and poisons (Clarke's isolation and identification of drugs). 3rd ed. London: Pharmaceutical Press; 2004, pp. 1–17.
[2] Uges DRA. www.bioanalysis.umcg.nl *(monthly up-dated)*

2.32 Novel aspects of the von Willebrand factor – platelet glycoprotein IB interaction and signaling

Ulrich Walter, Stepan Gambaryan, Sabine Herterich and Suzanne M. Lohmann

Summary

The major (but not only) roles of the hemostatic system are to keep blood in a fluid state under physiological conditions and to seal a vessel wall defect in order to prevent blood loss. Pathologically, bleeding or thrombosis may occur if the hemostatic stimulus is unregulated, either at the level of the stimulatory or inhibitory pathways. The balance of stimulatory [i.e., von Willebrand factor, adenosine diphosphate (ADP) and thrombin] and inhibitory [prostacyclin/cyclic AMP; nitric oxide (NO)/cyclic GMP] pathways is also well-established for platelets. These specialized adhesive cells play a key role in normal and pathological hemostasis through their ability to rapidly adhere to the subendothelial matrix proteins and to other activated platelets. Such platelet functions are strongly inhibited by endothelium-derived prostacyclin and NO and their cyclic AMP/protein kinase A (PKA)- and cyclic GMP/protein kinase G (PKG)-regulated pathways including their protein kinase substrates such as vasodilator-stimulated protein (VASP), inositol-1,4,5-triphosphate receptor-associated cGMP kinase substrate (IRAG), LIM and SH3 protein (LASP) others [1]. It is also well established that both stimulatory and inhibitory platelet pathways interact at various levels, representing significant functional cross-talk. Recently, we (re)addressed the interaction of the von Willebrand factor – gylcoprotein Ib (GPIb) pathway with the NO/cyclic GMP pathway in murine and human platelets. We discovered that human and murine platelets do not express functional nitric oxide synthase (NOS) proteins and that platelet-soluble guanylyl cyclase is NOS-independently activated by the von Willebrand factor, which may represent a new mechanism of feedback inhibition [2]. The underlying mechanisms are currently being studied using a recently-established quantitative phosphoproteomic approach [3]. Some recent data from this approach will be presented. As long-term goal, it is hoped that such studies will establish new diagnostic and perhaps also therapeutic approaches with respect to platelet function and dysfunction, especially when the NO/cyclic GMP signaling pathway is impaired.

References

[1] Walter U, Gambaryan S. cGMP and cGMP-dependent protein kinase in platelets and blood cells. Handb Exp Pharmacol 2009;191:533–48.
[2] Gambaryan S, Kobsar A, Hartmann S, Birschmann I, Kuhlencordt P, Müller-Esterl W, Lohmann SM, Walter U. NO-synthase-/NO-independent regulation of human and murine platelet soluble guanylyl cyclase activity. J Thromb Haemost 2008;6:1376–84.
[3] Lewandrowski U, Wortelkamp S, Lohrig K, Zahedi RP, Wolters DA, Walter U, Sickmann A. Platelet membrane proteomics: a novel repository for functional research. Blood 2009;114:e10–9.

2.33 Medical emergencies: what is the laboratory's role?

Donald S. Young

Summary

Laboratory tests play an essential role in the diagnosis, determination of treatment and disposition of patients in the emergency department. Emergency department physicians work under stressful conditions and must be expert in multiple branches of medicine. A large proportion of the population seeks medical care through the emergency department. Laboratory tests for such patients must be accurate and delivered in a timely manner, since there are few opportunities to repeat tests because of the need to process patients rapidly and appropriately.

The intent of the laboratory in medical emergencies

Improvements in medical care have been greatly assisted by the availability of high-quality laboratory tests. This is most evident in the emergency department (ED) of a hospital where laboratory test results are required for many critical decisions regarding the diagnosis and management of patients. The population visiting an ED tends to be heterogeneous, ranging from people with cuts or bruises to those who have experienced severe trauma. They also comprise individuals without physicians, and patients recently discharged from hospital. Overall, in the US, almost 40% of the population visits an ED each year. The percentage of infants attending the ED is 88.5% and in people aged 75 years or more is 62.0% = [1]. Although the reasons patients have for visiting EDs are varied, the most common symptom in patients visiting an ED is pain, which provides a challenge for differential diagnoses, although the frequency of any single reason for visiting an ED is typically <3%. The severity of conditions seen in the ED is such that as many of 24% of adults below 65 years of age and 47% of those over 65 years are admitted to hospital, although a critical analysis of such admissions shows that many were avoidable [2].

Laboratory professionals need to recognize that physicians in EDs have to be skilled in many branches of medicine and are extremely busy with many interruptions and distractions [3]. These physicians are very dependent on laboratory test results for decision-making. They are concerned about the selection of tests available to them at all times and the timeliness of the results. Laboratories have an obligation to ensure that results of the tests ordered are of high-quality and delivered to the ordering physician. Tests that are essential for ED physicians are shown in ►Tab. 2.33.1.

Analysis of actual usage shows that the most frequently ordered tests [1] are:

- complete blood counts, performed in 35.4% of all ED patients;
- urinalysis performed in 22.5%;
- urea and/or creatinine in 22.1%; and
- glucose in 19.6%.

Tab. 2.33.1: Essential tests for emergency departments.

Essential tests for emergency departments	
• Cardiac biomarkers	• Blood gases
• Glucose	• Liver function tests
• Electrolytes	• Amylase or lipase
• Creatinine and/or urea	• Toxicology screen
• Complete blood count	• Blood culture
• Prothrombin time/INR	• Urine pregnancy tests
• Dipstick urinalysis	

INR, International normalized ratio.

Regrettably, there is a high rate of failure to follow-up test results requested on patients in EDs. This is the case in up to 75% of tests in up to 16.5% patients [4].

When patients are admitted to hospital from EDs, the most common conditions are cardiac related: troponin I or T measurements play an important role in the decision to admit the patients. It is important that an appropriately sensitive cutoff point is used to determine that a patient has acute coronary syndrome, otherwise cases will be missed [5].

Up to 17% of medical patients are readmitted to hospital within 30 days of previous discharge and almost half of these occur within 10 days of discharge [6]. Many of the readmitted patients are on high-risk medications, such as steroids or opiates, and more than a quarter of the readmitted patients have same diagnosis as that associated with their previous hospitalization.

Urinalyses are the essential determinants of whether patients have urinary tract infections and complete blood counts are required to determine whether a patient is anemic. By performing a differential leukocyte count, the cause of a fever may be narrowed down. Since the time to diagnosis of a patient's illness and initiation of proper treatment should be minimized, the laboratory should provide its support to the ED through pneumatic tube delivery of specimens to an efficient central laboratory using prioritized testing for specimens from the ED, or through point-of-care testing for which the laboratory assumes responsibility for the quality of the results. If point-of-care testing is performed, results need to be comparable with regard to analytic sensitivity and specificity to those performed in a central laboratory. The laboratory staff should be prepared to respond rapidly to any unusual physician request that would enable him/her to make a correct diagnostic or treatment decision.

Conclusion

The laboratory carries out decision-making tests for ED physicians. The tests must be performed in a timely manner and be accurate. Whether or not the tests are performed in a central laboratory, the laboratory staff must assume responsibility for all aspects of their quality. The laboratory staff should seize the opportunities to interact with physicians in the ED to ensure optimal management of their patients.

References

[1] Niska R, Bhuiya F, Xu J. National hospital ambulatory medical care survey: 2007 Emergency department summary. National health statistics reports: no 26. Hyattsville, MD: National Center for Health Statistics; 2010.

[2] Ramos RA, Nafday A, Condon CJ, Thiessen D. Emergency department utilization in Orange County (2009). Orange County Health Care Agency, QM Research and Planning: Santa Ana, California; 2009.

[3] Chisholm CD, Weaver CS, Whenmouth L, Giles B. A task analysis of emergency physician activities in academic and community settings. Ann Emerg Med 2011;58:117–22.

[4] Callen J, Georgiou A, Li J, Westbrook JI. The safety implications of missed test results for hospitalised patients: a systematic review. BMJ Qual Saf 2011;20:194–9.

[5] Mills NL, Churchhouse AMD, Lee KK, Anand A, Gamble D, Shah ASV, Paterson E, et al. Implementation of a sensitive troponin I assay and risk of recurrent myocardial infarction and death in patients with suspected acute coronary syndrome. J Am Med Assoc 2011;305:1210–6.

[6] Allaudeen N, Vidyarthi A, Maselli J, Auerbach A. Redefining readmission risk factors for general medicine patients. J Hosp Med 2011;6:54–60.

www.ingramcontent.com/pod-product-compliance
Lightning Source LLC
Chambersburg PA
CBHW081225190326
41458CB00016B/5678

* 9 7 8 3 1 1 0 2 2 4 6 3 4 *